ASTRONOMY FOR THE INQUIRING MIND

ASTRONOMY
For the Inquiring Mind

THE GROWTH AND USE
OF THEORY IN SCIENCE

BY

ERIC M. ROGERS

PRINCETON UNIVERSITY PRESS

PRINCETON, NEW JERSEY

TO JANET TRAN ROGERS

FOREWORD

THIS IS the Astronomy section of my large physics-teaching text *Physics for the Inquiring Mind*, reprinted in this book for the same special purpose. Astronomy is part of physical science, but I do not use it here as one more field of knowledge to learn. It is here for a very important purpose: *to provide a clear example of the growth and use of THEORY in science.* With that aim this section deals with the history of our knowledge of the solar system—Sun, Moon, Earth, and planets—from early watching and simple fables to the magnificent success of Newton's gravitational theory. By giving a simple historical treatment we can make the nature of theory clearer than by describing it ready-made. A genuine understanding of theory and its relationship with experiment is essential if one wishes to know science; and gravitational planetary theory gives the best example here because readers need not just learn its results but can see why it is needed and how it is formed.

It would be a great further advantage if this history could also show the interplay between scientific discovery and the social environment, and the reactions between scientific theory and other branches of philosophy. However, that would require far more discussion of historical background. So this section is not a fair setting-forth of history. If accounts of people show a one-sided tendency to moralize, or if a few great scientists seem to stand like isolated lighthouses in a formless ocean, remember that this is not fair history, but only a restricted summary for a special purpose.

These chapters were published in the main text two decades ago. In the course of time the fabric of descriptions and explanations has developed a frayed edge here and there—also new discoveries have offered embroidery to add decoration. But to make corrections that would bring the material up to date would not only increase printing costs but would also probably add new frayed edges by changes of style. So, since I intend the present book to show *development of theory* rather than give up-to-date *information*, I have kept to the original text with only a few necessary corrections and some condensation.

In the chapters of *Physics for the Inquiring Mind* that follow the astronomy section (400 pages, more than half the book), readers will find other uses of theory as well as experiment in developing our knowledge of energy, electricity, atoms, nuclei. . . .

In the present book I have used the same figure numbers for illustrations as in the large text; so they serve as useful landmarks common to both books.

ERIC M. ROGERS
Princeton, New Jersey
May, 1981

SIGNS AND ABBREVIATIONS

$\times$ in $5 \times 2 = 10$ means "multiplied by"

$\cdot$ in $5 \cdot 2 = 10$
or in force $\cdot$ time
means "multiplied by"

$\approx$ means "equals approximately"

$\sim$ means "equals roughly"

$\propto$ means "is (or are) proportional to"
or "varies as"

$\triangle$ means "change of"

CONTENTS

〜〜〜〜〜〜〜〜〜〜〜〜〜〜〜〜〜〜〜〜

"Give me matter and motion, and I will construct the universe."
—René Descartes (1640)

"... from the phenomena of motions to investigate the forces of nature, and then from these forces to demonstrate the other phenomena; ... the motions of the planets, the comets, the moon and the sea. ..."
—Isaac Newton (1686)

"No one must think that Newton's great creation can be overthrown by [Relativity] or any other theory. His clear and wide ideas will forever retain their significance as the foundation on which our modern conceptions of physics have been built."
—Albert Einstein (1948)

PRELIMINARY INTRODUCTION

"What distinguishes the language of science from language as we ordinarily understand the word? How is it that scientific language is international? . . . The super-national character of scientific concepts and scientific language is due to the fact that they have been set up by the best brains of all countries and all times. In solitude and yet in cooperative effort as regards the final effect, they created the spiritual tools for the technical revolutions which have transformed the life of mankind in the last centuries. Their system of concepts has served as a guide in the bewildering chaos of perceptions, so that we learned to grasp general truths from particular observations."

—A. EINSTEIN, *Out of My Later Years*

Introduction

In this book you can study some methods of physical science and especially the development and use of some scientific theories. You will learn scientific facts and principles, some useful for life in general, others important groundwork for discussions. To gain much from your reading you need to learn this subject-matter carefully. In itself it may seem unimportant—such *factual* knowledge is easily forgotten,[1] and we are concerned with a more general understanding which will be of lasting value to you as an educated person—yet we shall use the factual knowledge as a means to important ends. The better your grasp of that knowledge, the greater your insight into the science behind it. And this book is concerned with the ways and work of science and scientists.

In this book we choose Astronomy for our series of samples of stages because it provides an unusually clear view of the growth of theories. We hope to lead readers to an understanding of the use of theories in science. However, to begin by discussing scientific methods or the structure of science would be like arguing about a foreign country before you have visited it. So we shall start with an easy sample of physics—gravity and falling objects —and then begin Astronomy.

What to Do about Footnotes

You are advised to read a chapter straight through first, omitting the footnotes. Then reread carefully, studying both text and footnotes. Some of the footnotes are trivial, but many contain important comments relevant to the work of the course. They are not minor details put there with a twinge of conscience to avoid their being omitted altogether. They

are moved out to make the text more continuous for a first reading. Often the footnotes wander off on a side issue and would distract attention if placed in the main text. Yet this developing of new threads itself shows the complex texture of scientific work; so at a second reading you should include the footnotes.

Falling Bodies

Watch a falling stone and reflect on man's knowledge of falling objects. What knowledge have we? How did we obtain it? How is it codified into laws that are clearly remembered and easily used? What use is it? Why do we value scientific knowledge in the form of laws? Try the following experiment before you read further. Take two stones (or books or coins) of different sizes. Feel how much heavier the larger one is. Imagine how much faster it will fall if the two are released together. You might well expect them to fall with speeds proportional to their weights: a two-ounce stone twice as fast as a one-ounce stone. Now hold them high and release them together. . . . Which are you going to believe: what you saw, what you expected, or "what the book says"?

People must have noticed thousands of years ago that most things fall faster and faster—and that some do not. Yet they did not bother to find out carefully just *how* things fall. Why should primitive people want to find out how or why? If they speculated at all about causes or explanations, they were easily led by superstitious fear to ideas of good and evil spirits. We can imagine how such people living a dangerous life would classify most normal occurrences as "good" and many unusual ones as "bad"—today we use "natural" as a term of praise and "unnatural" with a flavor of dislike.

This liking for the usual seems wise: a haphazard unregulated world would be an insecure one to live

---

[1] Once learned, it is easily relearned if needed later. Much of the difficulty of learning a piece of physics lies in understanding its background. When you understand what physics is driving at, the rules or calculations will seem sensible and easy.

in. Children emerge from the sheltered life of a baby into a hard unrelenting world where brick walls make bruises, hot stoves make blisters. They want a secure well-ordered world, bound by definite rules, so they are glad to have its quirky behavior "explained" by reassuring statements. The pattern of seeking security in order, which we find in growing children today, probably applied to the slower growing-up of primitive savages into civilized men. As civilization developed, the great thinkers codified the world—inanimate nature and living things and even the thoughts of man—into sets of rules and reasons. Why they did this is a difficult question. Perhaps some were acting as priests and teachers for their simpler brethren. Perhaps others were driven by childish curiosity—again a need to know definitely, born of a sense of insecurity. Still others may have been inspired by some deeper senses of curiosity and enjoyment of thinking—senses rooted in intellectual delight rather than fear—and these men might be called true philosophers and scientists.

You yourself in growing up run through many stages of knowledge, from superstitious nonsense to scientific sense. What stage have you reached in the simple matter of knowledge of falling objects? Check your present knowledge by actually watching some things fall. Take two different stones (or coins) and let them fall, starting together. Then start them again together, this time throwing both outward horizontally (Fig. 1-1). Then throw one outward

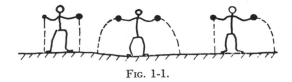

FIG. 1-1.

and at the same instant release the other to fall vertically. Watch these motions again and again. See how much information about nature you can extract from such trials. If this seems a childish waste of time, consider the following comments:

(i) This *is* experimenting. All science is built with information from direct experiments like yours.

(ii) To physicists the experiment of dropping light and heavy stones together is not just a fable of history; it shows an amazing simple fact that is a delight to see again and again. The physicist who does not enjoy watching a dime and a quarter drop together has no heart.

(iii) In the observed behavior of falling objects and projectiles lies the germ of a great scientific notion: the idea of *fields of force*, which plays an essential part in the development of modern mechanics in the theory of relativity.

(iv) And here is the practical taunt: if you use all your ingenuity and only household apparatus to try every relevant experiment you can think of, you will still miss some of the possible discoveries; this field of investigation is so wide and so rich that a neighbor with similar apparatus will find out something you have missed.

Mankind, of course, did not gather knowledge this way. Men did not say, "We will go into the laboratory and do experiments." The experimenting was done in daily life as they learned trades or developed new machines. You have been doing experiments of a sort all your life. When you were a baby, your bathtub and toys were the apparatus of your first physics laboratory. You made good use of them in learning about the real world; but rather poor use in extracting organized scientific knowledge. For instance, did your toys teach you what you have now learned by experimenting on falling objects?

Out of man's growing-up came some knowledge and some prejudices. Out of the secret traditions of craftsmen came organized knowledge of nature, taught with authority and preserved in prized books. That was the beginning of reliable science. If you experimented on falling objects you should have extracted some scientific knowledge. You found that the small stone and the big one, released together, fall together.[2] So do lumps of lead, gold, iron, glass, etc., of many sizes. From such experiments we infer a simple general rule: *the motion of free fall is universally the same, independent of size and material.* This is a remarkable, simple fact which people find surprising—in fact, some will not believe it when they are told it,[3] but yet are reluctant to try a simple experiment.[4]

[2] Yes; if you did not try the experiment, you now know the result of at least part of it. This is true of a book like this: by reading ahead you can find the answers to questions you are asked to solve. When you work on a crossword puzzle you would feel foolish to solve it by looking at the answers. In reading a detective story, is it much fun to turn to the end at once? Here you lose more still if you skip: you not only spoil the puzzle, but you lose a sense of the reality of science; you damage your own education. It is still not too late. If you have not tried the experiments, try them now. Drop a dime and a quarter together, and watch them fall. You are watching a great piece of simplicity in the structure of nature.

[3] Notice your own reaction to this statement: "A heavy boy and a light boy start coasting down a hill together on equal bicycles. In a short run they will reach the bottom together." The statement is based on the same general behavior of nature. See a demonstration. In a *long* run they gain high speeds and air resistance makes a difference.

The result *is* surprising. Would you expect a 2-pound stone to fall just as fast as a 10-pound one? Wouldn't it seem more reasonable for the 10-pound one to fall five times as fast? Yet direct trial shows that 1-pound, 2-pound, and 10-pound lumps of metal, stone, etc., all fall with the same motion.

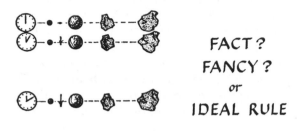

FACT ?

FANCY ?

or

IDEAL RULE

FIG. 1-2. FREE FALL

Early Science of Falling Bodies

What is the history of this piece of scientific knowledge? There may have been a long gap between casual observation and careful experimenting. Interest in falling objects and projectiles grew with the development of weapons. Spears, arrows, catapults, and more ambitious "engines of war" favored a simple vague knowledge of ballistics. But that took the form of craftsmen's working rules rather than scientists' understanding—unspoken familiarity rather than extracted simplicity. Two thousand years ago the Greeks thought and wrote about nature with genuine scientific interest, possibly inspired by similar activity earlier still in Egypt and Babylon. They gave rules and reasons for falling, but not very reliable ones. Though some of the ancient scientists must have experimented sensibly with falling objects, medieval use of the Greek written tradition set down by Aristotle (~ 340 B.C.) clouded the matter rather than clarified it, and led to a muddle which lasted for many centuries. Gunpowder greatly increased the interest in projectiles, but early cannon were still chiefly used to frighten the enemy when Galileo (~ 1600) rewrote the science of ballistics in clear rules that agreed with experiment. Those were rules for heavy slow cannonballs that ignored air resistance. Since then, speeding up of projectiles

has made air resistance more and more important, requiring modification of Galileo's simple treatment.

Aristotle and Philosophy

The great Greek philosopher and scientist Aristotle appears to have supported the popular idea that heavy things fall faster than light ones. Aristotle was a pupil of Plato and for a time the tutor of Alexander the Great. He founded a great school of philosophy and wrote many books. His writings were the authoritative sources of learning for centuries—through the dark ages when there were still no printed books but only handwritten ones copied and handed down by devout scholars in a rough and troubled world.

Why should philosophers be concerned with science? How is science related to philosophy? What *is* philosophy? Philosophy is not a weird high-brow scheme of impractical argument; it is *man's thinking about his own thoughts and knowledge*. Professional philosophy consists of criticizing knowledge, evolving systems of knowledge with rules of logic for critical argument. Philosophers are interested in questions of truth and nonsense, right and wrong, and in judgments of values. Just as professional physicians advise us on health, eating, sleeping, etc., so professional philosophers offer us advice on thinking and understanding, on all our intellectual activities. You and I indulge in amateur philosophy when we think intelligently about our life and its relation to the world around us, whenever we ask questions such as: "Is this really true?" "Does that really exist?" "What does it mean when I say something is right?" "Why is arithmetic right?" "Is happiness real or imaginary?" "Does a pin cause pain in the same sense that it causes a puncture?" Thinking about our place in the world is closely tied with our scientific knowledge of the world, so it is not surprising that the great philosophers studied science and influenced its progress. You cannot embark on science without a first step in philosophy. You need a philosophical assumption and a philosophical interest: you practically have to assume that there is an external world; and you have to wish to find out about it and "understand" it. And when you collect facts, formulate scientific laws, or invent theories, the philosopher in you will ask, "Are these *true*?" As you brood on that question you may change your view of science. When you have read this book, you may not have settled a general philosophy, but you will have done some philosophical thinking and you will have started making your own Philosophy of Science.

---

[4] We all rely in many matters on authority embedded in home teaching or common sense; we are reluctant to risk disturbing our sense of security. If you do not believe this accusation, wait and you will presently catch yourself.

Aristotle inherited a general philosophical. viewpoint from Plato. In trying to answer the question of ultimate truth and reality, Plato sheared away individual differences among the things we observe and extracted simple ideal forms. From dogs he extracted an ideal class, DOG; from all varieties of stones, an ideal STONE, and so on. Then he set forth the view that only these primary types or ideal forms really exist. These forms or essences remain universal and unchangeable, and individual examples of them are just shadows of the ideal. Aristotle used this insistence on classes of things as a basis for logical argument (If . . . , then . . . .). Yet, as a great observer and classifier of nature, he had to credit individual stones and dogs with some real existence; so his outlook was a compromise. Later students of his work gave increasing reality to ordinary objects, and came to treat the underlying classes as mental concepts or even mere names. This later view, that individual things are important and real, is a comfortable one for a scientist experimenting on objects and events in nature—he would like to feel that he is working with real things. Sir William Dampier seems to call some such view "nominalism" in the passage below, though modern philosophers use this name somewhat differently.

"Whatever be the truth of Plato's doctrine of ideas from a metaphysical point of view, the mental attitude which gave it birth is not adapted to further the cause of experimental science. It seems clear that, while philosophy still exerted a predominating influence on science, nominalism, whether conscious or unconscious, was more favourable to the growth of scientific methods. But Plato's search for the 'forms of intelligible things' may perhaps be regarded as a guess about the causes of visible phenomena. Science, we have now come to understand, cannot deal with ultimate reality; it can only draw a picture of nature as seen by the human mind. Our ideas are in a sense real in that ideal picture world, but the individual things represented are pictures and not realities. Hence it may prove that a modern form of [Plato's view of] ideas may be nearer the truth than is a crude nominalism. Nevertheless, the rough-and-ready suppositions which underlie most experiments assume that individual things are real, and most men of science talk nominalism without knowing it . . .

"The characteristic weakness of the inductive sciences among the Greeks is explicable when we examine their procedure. Aristotle, while dealing skilfully with the theory of the passage from particular instances to general propositions, in practice often failed lamentably. Taking the few available facts, he would rush at once to the widest generalizations. Naturally he failed. Enough facts were not available, and there was no adequate scientific background into which they could be fitted. Moreover, Aristotle regarded this work of induction as merely a necessary preliminary to true science of the deductive type, which, by logical reasoning, deduces consequences from the premises reached by the former process."[5]

While Aristotle may be regarded as having given *experimental* science a strong push forward, Plato was perhaps nearer to the modern *theoretical* physicist with his insistence on the importance of underlying general forms and principles. As a tool for his thinking, Aristotle developed a magnificent system of formal logic, that is, cast-iron argument that starts from admitted facts or agreed assumptions and draws a compelling conclusion. In treating science he first tried to extract some general scientific principles from observations, a process we call *induction*. Then he reasoned logically from these principles to deduce new scientific knowledge. His system of logic was itself a magnificent discovery, but it cramped the development of early experimental science by directing too much attention to argument. It has influenced the growth of our civilization profoundly. Most of us never realize how much our pattern of thinking has been influenced by the age-long tradition of Aristotelian logic, though many thinkers today question its rigid simplicity. It argued from one absolute yes or no to another absolute yes or no; argued with good logic to a valid conclusion, provided the starting-point was valid. "Is every man mortal?" "Do 4 times 3 make 14?" "Do 2 plus 2 make 4?" "Do all dogs have 7 legs?" We answer any of these with an absolute "yes" or "no" and then deduce answers to questions such as, "Is Jones mortal?" "Does my terrier have 7 legs?" But try the following:

"Is self-sacrifice good?"

"Was Lincoln a success?"

"Is my Boyle's law experiment right?"

These are important questions, but we can make fools of ourselves by insisting on a yes or no answer. If instead we spread our judgments over a wider scale of values, we may lose some "logic" but gain greatly in intellectual stature. It is well to beware of people who try to dissect every problem or dis-

[5] Sir William Dampier in *A History of Science* (4th edn., Cambridge University Press, Cambridge, Eng., 1949), pp. 34-35, from which some of this discussion is drawn.

cussion into components that have an absolute yes or no.

Aristotle's logic was safe as far as it went; modern logicians regard it as restricted and unfruitful but "true."[6] The damage to your thinking and mine comes from centuries of medieval scholarship drawing blindly and insistently on his writings—"the ingrown, argumentative, book-learned, world-ignorant atmosphere of medieval university learning." That medieval Aristotelian tradition is built into today's language and thought, and people often mistakenly require an absolute yes or no. For example, people trained to think they must choose between complete success and complete failure are heartbroken when they find they cannot attain the impossible goal of complete success. We are all in danger: students in college, athletes in contests, men and women in their careers, older people reviewing their life—all face terrible discouragement or worse if they demand absolute success as the only alternative to failure. Fortunately, many of us achieve a wiser balance; we stop judging ourselves by an absolute yes or no and enjoy our own measure of success. We then find the conflicting mixture of our record easier to live with.[7]

In science, where simple logic once seemed so safe, we are now more careful. Asked whether a beam of light is a wave, we no longer assume there is an absolute yes or no. We have to say that in some respects it is a wave and in others it is not. We are more cautious about our wording. Remembering that our modern scientific theory is more a way of regarding and understanding nature than a true portrait of it, we change our question, "*Is* it a wave?" to "Does it *behave as* a wave?" And then we can answer, "In some circumstances it does, in other circumstances it does not." Where an Aristotelian would say an electron must be either inside a certain box or not inside it, we have to say we would rather regard it as both! If you find such cautiousness irritating and paradoxical, remember two things: first, you have been brought up in the Aristotelian tradition (and perhaps you would be wise to question its strong authority); second, physicists themselves shared your dismay when experiments first forced some changes of view on them, but they would rather be true to experiment than loyal to a formality of logic.

Aristotle and Authority

Aristotle's chief interests lay in philosophy and logic, but he also wrote scientific treatises, summing up the knowledge available in his day, some 2000 years ago. His works on Biology were good because they were primarily descriptive. In his works on Physics he was too much concerned with laying down the law and then arguing "logically" from it. He and his followers wanted to explain *why* things happen and they did not always bother to observe *what* happens or *how* things happen. Aristotle explained why things fall quite simply: they seek their *natural place*, on the ground. In describing how things fall, he made statements such as these: ". . . just as the downward movement of a mass of lead or gold or of any other body endowed with weight is quicker in proportion to its size. . . ." ". . . a body is heavier than another which in equal bulk moves down more quickly. . . ." He was a very able man, discussing as a philosopher the *why* of falling, and he probably had in mind a more general survey of falling bodies, knowing that stones do fall faster than feathers, blocks of wood faster than sawdust. In the course of a long fall air-friction brings a falling body to a steady speed, and he probably referred to that.[8] But later generations of thinkers and teachers who used his books took his statements baldly and taught that "bodies fall with speeds proportional to their weights."

The philosophers of the Middle Ages grew more and more concerned with argument and disdained experimental tests. Most of the earlier writings on geometry and algebra had been lost and experimental physics had to wait until they were found and translated. For centuries, right on through the "dark ages," the authority of Aristotle's writings

[6] Roughly speaking, Aristotelian logic deals with classes of things, and its arguments can be carried out by machines, e.g., "electronic brains" in which "yes" or "no" is signified by an electron stream being switched "on" or "off." Modern logic deals with *relations* (such as ". . . larger than . . . ," ". . . better than . . .") as well as classes (such as "dogs," "mammals") and, nowadays, with implicational relations between complete propositions. Its arguments, too, can probably be carried out by machines, though that may be more difficult to arrange. But a machine cannot criticize the system of logic that it is asked to administer. Only man still thinks he can do that, making judgments of value.

For descriptions of machines see the following numbers of *Scientific American*:

Vol. 183, No. 5, "Simple Simon" (a small mechanical brain); Vol. 180, No. 4, "Mathematical Machines" (a detailed account of electronic calculators); Vol. 182, No. 5, "An Imitation of Life" (mechanical animals that learn); Vol. 185, No. 3, "Logic Machines"; Vol. 192, No. 4, "Man Viewed as a Machine" (excellent article by a philosopher); Vol. 197, No. 3, a complete issue on self-regulating machines.

[7] For a fuller discussion, see Ch. I of *People in Quandaries* by Wendell Johnson (Harper and Bros., New York, 1946).

[8] A denser body (or a bigger one) has to fall farther before approaching its limiting speed; and then that speed is much greater.

remained supreme, in a misinterpreted form at that. Simple people, like children, love security more than freedom; they will worship authority blindly, and swallow its teaching whole. You may smile at this and say, "We are civilized. We don't behave like that." But you may presently ask, "Why doesn't this book give us the facts and tell us the right laws, so that we can learn reliable science quickly?"; and that would be *your* demand for simple authority and easy security! We now condemn "Aristotelian dogmatism" as unscientific, yet there are still people who would rather argue from a book than go out and find what really does happen. The modern scientist is realistic; he tries experiments and abides by what he gets, even if it is not what he expected.

Logic and Modern Science

Wholesale appeal to Aristotle's logic may restrict our intellectual outlook, and medieval wrangling with it certainly hampered science; but logic itself is an essential tool of all good science. We have to reason inductively, as Aristotle did, from experiments to simple rules. Then we often assume such rules hold generally and reason deductively from them to predictions and explanations. Some of our reasoning is done in the shorthand logic of algebra, some of it follows the rules of formal logic, and some of it is argued more loosely.

In extracting scientific rules from old laws we trust the "Uniformity of Nature": we trust that what happens on Friday and Saturday will also happen on Sunday; or that a simple rule which holds for several different spiral springs will hold for other springs.[9] Above all, we rely on the agreement of other observers. That is what makes the difference between dreams and hallucinations on one hand and science on the other. Dreams belong to each of us alone, but scientific observations are common to many observers. In fact, scientists often refuse to accept a discovery until other experimenters have confirmed it.

Scientists do more than assume that nature is simple, that there are rules to be found; they also assume that they can apply logic to nature's ways. There lies the essential distinction that enabled science to emerge from superstition: a growing belief that *nature is reasonable*. As science grows, mathematics and simple logic play an essential part as

[9] The obvious condition, "all other circumstances remain the same," is often difficult to maintain, and we blame many an exception to the Uniformity of Nature on some failure in that respect. Magnetic experiments in towns that have streetcars may give different results on Sundays, when fewer cars are running.

faithful servants. The modern scientist puts them to more use than ever, but he goes back to nature for experimental checks. In a sense, the ideal scientist has his head in the clouds of speculation, his arms wielding the tools of mathematics and logic, and his feet on the ground of experimental fact.

Greeks to Galileo

"In studying the science of the past, students very easily make the mistake of thinking that people who lived in earlier times were rather more stupid than they are now."—I. BERNARD COHEN

Aristotle's authority grew and lasted until the 17th century when the Italian scientist Galileo attacked it with open ridicule. Meanwhile, many people must have privately doubted the Aristotelian views on gravity and motion. In the 14th century a group of philosophers in Paris revolted against traditional mechanics and devised a much more sensible scheme which was handed down and spread to Italy and influenced Galileo two centuries later. They talked of *accelerated motion*, and even of *uniform acceleration* (under archaic names), and they endowed moving objects with "impetus," meaning motion or momentum of their own, to carry them along without needing a force.

Galileo ($\sim$ 1600) was a great scientist. He started science advancing to a new level where critical thinking and imagination join with an experimental attitude—a partnership of theory and experiment. He gathered the available knowledge and ideas, subjected them to ruthless examination by thinking, experimenting, and arguing; and then taught and wrote what he believed to be true. He lost his temper with the Aristotelians when they disliked his teaching and disdained his telescope; and he wrote a scathing attack on their whole system of science, setting forth his own realistic mechanics instead. He cleared away cobwebs of muddled thinking and built his scheme on real experiment—not always his own experiments, more often those of earlier workers whose results he collected.

Thought-Experiments

In his books and lectures, Galileo often reasoned by drawing on common sense, quoting "thought-experiments." For example, he discussed the breaking-strength of ropes in this manner: suppose a rope 1 inch in diameter can just support 3 tons. Then a rope of double diameter, 2 inches, has four times the cross-section area (πr^2) and therefore four times as many fibers. Therefore, the rope of double diameter has four times the strength—it should support 12 tons. In general, STRENGTH must increase as DIAMETER². Galileo gave this argument and extended it to wooden beams, pillars, and bones of animals.[10] Some thought-experiments deal with simplified or idealized conditions, such as an object falling *in a vacuum.*[11]

Ideal Rules for Free Fall

Galileo realized that air resistance had entangled the Aristotelians. He pointed out that dense objects for which air resistance is relatively unimportant fall almost together. He wrote: ". . . the variation of speed in air between balls of gold, lead, copper, porphyry, and other heavy materials is so slight that in a fall of one hundred cubits a ball of gold would surely not outstrip one of copper by as much as four fingers. Having observed this, I came to the conclusion that, in a medium totally devoid of all resistance, all bodies would fall with the same speed."[12] By guessing what would happen in the imaginary case of objects falling freely in a vacuum, Galileo extracted ideal rules:

(1) All falling bodies fall with the same motion; started together, they fall together.

(2) The motion is one "with constant acceleration": the body gains speed at a steady rate; it gains the same addition of speed in each successive second.

Having guessed the rules for the ideal case, he could test them in real experiments by making allowances for the complications of friction.

Galileo's ? Experiment: Myth and Symbol

There is a fable that Galileo gave a great demonstration of dropping a light object and a heavy one from the top of the leaning tower of Pisa.[13] (Some say he dropped a steel ball and a wooden one, others say a 1-pound iron ball and a 100-pound iron ball.)

[10] See problems in Chapter 5.
[11] The Aristotelians had argued themselves into believing a vacuum to be impossible, so they cut themselves off from Galileo's satisfying simplification.
[12] From *Dialogues Concerning Two New Sciences*, by Galileo Galilei, English translation by H. Crew and H. de Salvio, Northwestern University Press, p. 72.
[13] Pisa. The leaning tower is a charming little building in a friendly Italian town. It is a round tower of white marble, built beside the cathedral. It began to lean as it was built, and it now has a remarkable tilt, about 5° from the vertical. The visitor who climbs its winding stair or walks around one of its open slanting balconies has strange sensations of shifting gravity. The tower was built long before Galileo's day, and he must have tried using it for some experiments. In his lifetime a pro-Aristotelian used the tower, to demonstrate *unequal* fall.

There is no record of such a public performance, and Galileo certainly would not have used it to show his ideal rule. He knew that the wooden ball would be left far behind the iron one, but he said that a taller tower would be needed to show a difference for two unequal iron balls. He certainly tried rough experiments as a youth and knew as you do what does happen, but he did not suddenly turn the course of science with one fabulous experiment. He did accelerate the growth of real physics by refuting the Aristotelians' silly dogmatic statements. And he did start science on a new kind of growth by applying his simplifying imagination to experimental knowledge. These, and not the leaning tower, made him a landmark in scientific history. Many a myth is attached to great figures in history—stories about cherry trees, burning cakes, etc. Though scholars delight in debunking these anecdotes, they also use some of them to show how the people of the great man's day thought of him. The leaning tower story is not even credited with that advantage. Yet we might use it, quite apart from Galileo and the growth of science, as a symbol of a simple experiment. In your own experiment with unequal stones, they fell almost together, and not, as some people

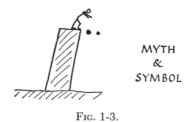

MYTH & SYMBOL

FIG. 1-3.

expect, the heavy one much faster. We shall use this Myth & Symbol in our course as a reminder of two things: the need for direct experiment, and a surprising, simple, important fact about gravity.

Honest Experimenting vs. Authority

Your own experiments do not show that all things fall together; they did not even show that large and small stones fall *exactly* together; and if in obedience to book or teacher you said, "They fall exactly together," you were cheating yourself of honest science. Small stones lag slightly behind big ones—the difference growing more noticeable the farther they fall. Nor is it simply a matter of different sizes: a wooden ball and a steel ball of the same size do not fall exactly together.

Once you accept Galileo's view that air resistance obscures a simple story, you can interpret your own observations easily—though that still leaves air resistance to be investigated. Or you might pretend you had never heard Galileo's view, and proceed towards it yourself through a series of experiments with denser and denser objects. Finding the motion more and more nearly the same for larger or denser bodies, you might guess the rule for the ideal case. To examine the blame against air resistance, you might try streamlining, or reducing, using some object such as a sheet of paper.

Galileo's Guess: Newton's Crucial Experiment

Galileo could only *decrease* air resistance. He could not remove it completely, so he had to argue from real observations with less and less resistance to an ideal case with none. This intellectual jump, from real observations to an ideal case, was his great contribution. Then, looking back, he could "explain" the differences in real experiments by blaming air resistance. He could even study air resistance, codify its behavior, and learn how to make allowances for it. Not long after his time, air pumps were developed which enabled people to try free-fall experiments in a vacuum. Newton pumped the air out of a tall glass pipe and released a feather and a gold coin at the top. Even this extreme pair fell together. *There* was a crucial test of Galileo's guess.

Scientific Explanations

When we "explain" the differences of fall by air resistance, the term "explain" means, as so often in science, to point out a likeness between the thing under investigation and something else already known. We are saying essentially, "You know about wind resistance, when you move a thing along in the air. Well, the falling bodies experience wind resistance which depends in some way on their bulkiness. A wooden ball and a lead ball of the same size moving at the same speed would suffer the same air resistance—how could the air know or mind what is inside?—but the lead ball weighs more, is pulled harder by gravity, *so the air resistance matters less to it in comparison with the pull of the Earth.*"[14]

Further Investigation

The explanation leads to a whole new line of enquiry: wind resistance, fluid friction, streamlining—with applications to ballistics and airplane design—new science from more accurate study of

[14] At this stage, the explanation ends in unsupported dogmatism that might be "straight out of Aristotle." Wait for studies of mass, force, and motion to make it good science.

some simple rule of behavior, from a study of its failures.

You could extend your series of experiments in the other direction, making more and more resistance, first with air, then with water, and find things of importance in the design of ships and planes. For simple experiments with fluid friction, try dropping small balls in water instead of in air. Balls of different sizes do not fall together. Moreover they fail to move any faster after a while in a long fall. Each ball seems to reach a fixed speed and then move steadily down at that speed. What is happening then? Investigations might lead you to Stokes' Law for fluid friction on a moving ball, a law which plays a vital part in measuring the electric charge of a single electron. If you investigated still smaller falling bodies, specks of dust or drops of mist, you would discover surprising *irregularities* in their fall, and these in turn could lead to useful information in atomic physics.

Galileo's experimenting and thinking, which you have been repeating, led to a simple rule that applies accurately to objects falling in a vacuum. For things falling in air, it applies with limited precision. In other words, the simple statement "all freely falling bodies fall together" is an artificial extract distilled by scientists out of the real happenings of nature. This is a good scientific procedure: first to extract a general rule, under simplified or restricted conditions, secondly to look for modifications or exceptions and then to use them to polish up the rule and to extend our knowledge to new things. In the case of falling bodies we can now test the extracted rule by dropping things in a vacuum. Ask to see Newton's "guinea and feather" experiment. In many cases in physics, however, we have to be content with knowing that our rule is an extracted simplification, believing in it as a sort of ideal statement, with only indirect evidence to justify our full belief.

Restricting the Number of Variables

Apart from ignoring air resistance, we have restricted our study of falling bodies in another way: we have concentrated attention on just one aspect of them, their comparative rate of fall. We have not observed what noise they make as they fall, or watched how they spin, or looked for temperature changes, etc. By narrowing our interests, temporarily, we have better hopes of finding a simple guiding rule. Again this is good scientific procedure. In many investigations we not only concentrate on a few aspects but even arrange to hold other aspects constant so that they do not muddle the investigation. In physics we nearly always try to limit our investigation to one pair of variables at a time. For example, we compress a sample of air and measure its VOLUME at various PRESSURES, while we *keep the temperature constant*. Or we warm up the gas and measure the PRESSURE at various TEMPERATURES, while we *keep the volume constant*. From these experiments we can extract two useful "gas laws" that can be combined into one grand law. If we did not make restrictions but let TEMPERATURE and PRESSURE and VOLUME change during our experiments, we could still discover the grand law but our measurements would seem mixed and complicated—it would be harder to see the simple relationship connecting them. But other sciences such as biology and psychology, following the successful example of physics, have found this method very dangerous. While restricting attention to one aspect of growth or behavior, the investigator may lose sight of the body or mind as a whole. In attempts to apply the methods of natural sciences to social sciences such as economics this danger is even more severe.

Why Do Things Fall?

Aristotle was concerned with the answer to "Why?" Why *do* things fall? What is *your* answer to the question? If you say, "because of gravitation or gravity," are you not just taking refuge behind a long word? These words come from the Latin for heavy or weighty. You are saying, "things fall because they are heavy." Why, then, are things heavy? If you reply "because of gravity," you are talking in a circle. If you answer, "because the Earth pulls them," the next question is, "How do you know the Earth goes on pulling them *when they are falling?*" Any attempt to demonstrate this with a weighing machine during fall leads to disaster. You may have to answer, "I know the Earth pulls them because they fall"—and there you are back at the beginning. Argument like this can reduce a young physicist to tears. In fact, physics does not explain gravitation; it cannot state a cause for it, though it can tell you some useful things about it. The Theory of General Relativity offers to let you look at gravitation in a new light but still states no ultimate cause. We may say that things fall because the Earth pulls them, but when we wish to explain why the Earth pulls things all we can really say is, "Well it just *does*. Nature *is* like that."[15] This is disappointing to people

who hope that science will explain everything, but we now consider such questions of ultimate cause outside the scope of science. They are in the province of philosophy and religion. Modern science asks *what?* and *how?* not the primary *why?* Scientists often explain why an event occurs, and you will be asked "why . . . ?" in this course; but that does not mean giving a first cause or ultimate explanation. It only means relating the event to other behavior already agreed to in our scientific knowledge. Science can give considerable reassurance and understanding by linking together seemingly different things. For example, while science can never tell us what electricity *is*, it can tell us that the boom of thunder and the crack of a man-made electric spark are much the same, thus removing one piece of fearful superstition.

Aristotle's explanation of falling was: "The natural place of things is on the ground, therefore they try to seek that place." People today call that a silly explanation. Yet it is in a way similar to our present attitude. He was just saying, "Things *do* fall. That's *natural*." He carried his scheme too far however. He explained why clouds float upward by saying that *their* natural place is up in the sky and thus he missed some simple discoveries of buoyancy.[16] Aristotle was much concerned with stating the "natural place" and "natural path" for things, and he distinguished between "natural motion" (of falling bodies) and "violent motion" (of projectiles). He might have produced good science of force and motion except for a mistake of applying common knowledge of horses pulling carts to all motion. If the horse exerts a constant pull, the cart keeps going with constant speed. This probably suggested Aristotle's general view that a constant force is needed to keep a body moving steadily; a larger force maintains larger speed in proportion. This is a sensible explanation for pulling things against an adjustable resisting force, but it is misleading for falling bodies and projectiles. In all cases it forgets the resisting force is there and prevents our seeing what happens when there is no resistance.

To explain the motion of projectiles, the Greeks

[15] Parents often give answers such as, "Well it just *is*" or "because it *is*" to children's questions. Such answers are not so foolish as they sound. For a child they provide the reply

that is really needed at that stage, an assurance that everything is normal, that the matter asked about is a part of a consistent world. When a child asks "Why is the grass green?" he does not want to have a lecture on chlorophyll. He merely wants to be reassured that it is o.k. for the grass to be green.

[16] Buoyancy affects falling objects. When a thing falls in water its effective weight is lessened by buoyancy, and this makes falling in water quite different for different objects. Even air buoyancy has some effect, trivial for cannon balls, overwhelming for balloons.

imagined a "rush of air" to keep them going; and even more mysterious agents were required to keep the stars and planets moving. On their view—shove is needed to maintain motion—an arrow was kept moving by the push of the bowstring until it left the bow. After that another pushing agent had to be invoked to keep the arrow moving. Aristotelian philosophers imagined a rush of air pushing the arrow, not just a gust of wind travelling with it, but a circulation of air, with the air ahead of the arrow being pushed aside and running around to shove the arrow from behind. This rush of air satisfactorily prevented the unthinkable vacuum forming behind the arrow.[17] So firmly established was the idea of a rush of air, with embellishments of initial commotions, that it was used as an argument to show that projectiles could not move in a vacuum. "In a vacuum with zero resistance any force would maintain infinite velocity," argued the Greeks, "therefore a vacuum is impossible." God could never make a vacuum. Aristotle himself understood that all things would fall equally in a vacuum, but he considered that too a proof that a vacuum could not exist.

Mass

Whatever gravity really is, falling bodies do fall together except for effects of air resistance. This hints at a useful idea which we shall meet again and again: the idea of mass. Suppose we have a 2-pound lump of lead and a 1-pound lump. When we hold them, we feel the big lump being pulled more; we feel its greater weight. That is why we expect it to fall faster. Yet it does not, so there must be some other factor involved, something that the doubled weight-pull has to contend with. The reason is: *there is twice as much lead to be moved.* The double chunk needs twice the pull to give the same motion to its double quantity of lead. Galileo felt his way towards this idea of quantity of matter, which we shall call mass, but it had to wait for Newton to state it clearly. Mass is not an easy idea to grasp, but we shall return to it many times, because it plays a very important part in physics. At this stage, the amazing thing about gravitation is this: whatever the materials, gravitational pulls are exactly proportional to the amounts of stuff being pulled. Gravity, the mysterious pulling agent, seems to be ready to pull indiscriminately on any body, whatever it is made of, ready to pull just twice as hard on two bricks as on one, four times as hard on

4 cubic feet of lead as on 1 cubic foot, so that the object with more stuff in it to be pulled on has just the bigger pull needed to make it fall with the same motion as a smaller object.

Gravitational Field

We give a name to this state of affairs all around us of gravity-prepared-to-pull. We say there is a *gravitational field.* We are not really explaining anything by inventing a new term,[18] but we shall find it useful later. For the moment you should think of a gravitational field as waiting to clutch (proportionally) on any piece of matter put there, to pull it down towards the Earth, to make it fall. Near a magnet there is a similar state of affairs for bits of iron, a *magnetic field* waiting to clutch them. In your television tube, *electric fields* and *magnetic fields* clutch the whizzing electrons, speed them up, and swing the beam to and fro to sketch the picture.

Here we have been letting our thoughts run away with new words and ideas, such as MASS and FIELD, arising from simple experimenting. If we just worship such new ideas and phrases we are liable to fall back to a state of witchcraft. But if we use them to develop our knowledge, and if we put our suggestions to experimental tests, they may help the progress of science.

Galileo's Argument

Galileo was a great arguer. The Aristotelians had woven a web of "scientific" arguments based on Aristotle's statements, but Galileo beat them with their own weapons. An argument would upset them more than an experimental demonstration. So he revived a thought-experiment which ran thus: Take three equal bricks, A, B, C. Release them together, to fall freely. Now chain A and B together (by an invisible chain which is not really there) so that they form one object A + B which is twice as heavy as C.

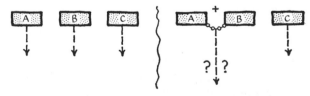

Fig. 1-4. Galileo's Thought-Experiment

Again release them. The Aristotelians would now expect A + B to fall twice as fast as C; yet since it

---

[17] For a fuller and very interesting account of these views on motion, see H. Butterfield, *The Origins of Modern Science* (New York, 1952), Ch. 1.

[18] Cf. the great use in psychology or biology of special words such as "repression," "complex," "heliotropism," etc. Such new terms, coined or adopted for scientific use, cannot explain things, but they can aid clear thinking and discussion.

is really only two separate bricks it will fall just as before, at the same rate as C. Therefore, the double brick A + B and the single one C must fall together. "Ah no," says the Aristotelian in the argument, "there is the chain that joins A and B. One of the bricks will somehow get a little ahead of the other and then it will drag the other downward, making the combine fall faster." "I see," says the Galilean spokesman, "then the other, being a little behind, drags the first one back, making the combine fall slower!" Can you see in the comparison of A + B and C, the germ of the idea of mass?

The Motion of Free Fall

If all freely falling bodies have the same motion, that motion itself is worth detailed investigation. It might tell us something about nature in general, something common to all falling things. We can see that falling bodies move faster and faster; they accelerate. (This is merely a word meaning "move faster"—using it does not make our statement more scientific.) Just what kind of accelerated motion do they have?

(1) Does the SPEED increase by sudden jumps? Experiment says no.

(2) Does the SPEED increase in direct proportion to the DISTANCE TRAVELLED? Galileo devised an ingenious argument to show that this is very unlikely.[19]

(3) Does the SPEED increase in direct proportion to the TIME?

(4) or to the (TIME)²,

(5) or in some other, more complicated manner?

Since we are asking a question about real nature, only experiments on real nature can answer it. (If you want to know how tall Abraham Lincoln was you must find out from someone who actually measured his height. Information from books is useless unless it came originally from real measurements. Algebra alone cannot possibly tell you.) We might go straight to the laboratory and experiment wildly and boldly, hoping to extract the essential story

[19] Galileo's argument was ingenious but not quite sound. It ran thus: "Compare two trips, each starting from rest, trip A of a certain distance, trip B of twice that distance. Then if speed increases in proportion to distance travelled, the speeds at corresponding stages (half way, ¾ way, etc.) of trip B are twice those of A. Then the double trip, B, is travelled with doubled speeds. Therefore B would take the same total time as A; which is absurd." But this argument supposes that the motion could start from rest. A sound version of the argument requires calculus to show that such a motion could never start from rest. Given a start, however, such a motion would continue in an ever increasing rush, its speed growing with compound interest.

from a host of measurements. Or we might do some thinking first, guess cunningly at some simple types of motion, calculate the consequences of each, and then go into the laboratory and experiment on the consequences. Both methods have contributed to the growth of science.

Inductive and Deductive Methods

The first method is named the *inductive* method. We gather information either in a laboratory or from the accumulated lore of some trade; then we extract from it some simple rule, or story about nature. We call this extracting process "inductive inference" or simply "induction." We first gather experimental data and then infer some general rule or scientific law from the data. For example, after watching the Moon for some years an observer might extract the general rule that the Moon travels around the Earth regularly, about 13 times a year, and it might seem a safe inference by induction that this will continue. Again, from an extensive record of eclipses, we might *infer inductively* a rule that eclipses of the Moon run in several regular series, with a fixed time-interval of about 18 years between successive eclipses in any one series.

The second method is named the *deductive* method. We start with some general rules or ideas then derive particular consequences or predictions from them by logical argument. If we are scientists, we then test the predictions experimentally. If experiment confirms the predictions, we continue our scheme. If it disagrees with our predictions, we throw doubt on our original assumptions and try to modify them. For example, we might assume that eclipses of the Moon are due to the Earth getting in the Sun's light, casting a shadow on it; assume simple orbital motions for Sun and Moon, and then *infer deductively* (or *deduce*) that an eclipse must occur again after an interval of time sufficient for Sun *and* Moon to return to the same positions relative to the Earth. This interval must be the "lowest common multiple" of one Moon-month and one Sun-year. So, by combining simple observations with sensible assumptions we could make a striking deductive prediction of the 18-year repeat cycle of eclipses. (For a successful calculation, the "Sun-year" must be a special, short year geared to the Moon's changing orbit.)

As Lancelot Hogben points out:

"Readers of crime fiction will be familiar with two types of detectives. One adopts the card index method of Francis Bacon, collecting all relevant information piece by piece. The other follows a hunch, like Newton, and, like Newton, abandons it at once when it comes into conflict with observed facts. From time to time the philosophers of science emphasize the merits of one or the other, and write as if one or the other were the true method of science. There is no one method of science. The unity of science resides in the nature of the result, the unity of theory with practice. Each type of detection has its use, and the best detective is one who combines both methods, letting his hunch lead him to test hypotheses and keeping alert to new facts while doing so."[20]

And here is an overall view, from a leading American physicist, P. W. Bridgman:

"I like to say that there is no scientific method as such, but that the most vital feature of the scientist's procedure has been merely to do his utmost with his mind, *no holds barred*."[21]

Accelerated Motion: Inductive and Deductive Treatment

Much of the *early* growth of science was made by induction; general laws were inferred from the knowledge gained in crafts and trades. In a simple way we have treated falling bodies inductively, inferring from many observations a general statement that all bodies falling freely in a vacuum fall together. When Galileo studied the details of this falling motion, he probably used a mixture of two approaches. He was good at making guesses, and he used geometry and reasoning powerfully.

We shall now follow the second method, deduction, in our study. We shall start by *assuming* a likely rule, and then we shall make a test comparing its consequences with real falling motion.

We choose guess (3) above and *assume that a falling body gains speed steadily, gaining equal amounts of speed in equal stretches of time.* We can express this more conveniently if we give a definite meaning to the word *acceleration*, so that we can say "the acceleration is constant." Therefore, we give the name ACCELERATION to

$$\frac{\text{GAIN OF SPEED}}{\text{TIME TAKEN}} \text{ or RATE-OF-CHANGE OF SPEED}$$

In making this definition of acceleration, we are really *choosing* the thing (GAIN OF SPEED)/(TIME

TAKEN) to work with, and then giving it a name. We are not discovering some true meaning which the word acceleration possessed all along! We make this choice and assign it a name because it turns out to be useful in describing nature easily.

We shall start using the grander word *velocity* instead of *speed*, and presently we shall make a distinction between their meanings. Since we shall often deal with changing things, we want a short way of writing "change of . . ." or "gain of . . . ." We choose the symbol Δ, a capital Greek letter D, pronounced "delta." It was originally used to stand for the *d* of "difference." Then our definition[22] of acceleration states that:

$$\text{ACCELERATION} = \frac{\text{GAIN OF VELOCITY}}{\text{TIME TAKEN}}$$
$$= \frac{\text{CHANGE OF VELOCITY}}{\text{CHANGE OF TIME-OF-DAY}}$$
$$a = \frac{\Delta v}{\Delta t}$$

where a, v, and t are obvious shorthand.

Deductive Treatment of Motion with Constant Acceleration

Now we express our assumption about falling bodies in this new terminology. We are *assuming* that:

$\frac{\Delta v}{\Delta t}$ is *constant*, for bodies freely falling (in

vacuum). This states a huge assumption regarding real nature. Is it true? Is $\Delta v/\Delta t$ constant? To test this directly we should need an accelerometer to measure the acceleration of a body, $\Delta v/\Delta t$, at each stage of its fall. Such instruments are manufactured, but they are complicated gadgets which would not provide convincing proofs at this stage. Instead, we follow Galileo's example and ask mathematics, the logical machine, to grind out a consequence of our assumption, and then we test the consequence by experiment. The machine tells us that:

IF the acceleration a ($= \Delta v/\Delta t$) is constant, and s is the distance travelled in time t with this constant acceleration, THEN

$s = \frac{1}{2}at^2$, if the motion starts from rest

$s = v_0 t + \frac{1}{2}at^2$, if the motion starts with velocity v_0 at the instant $t = 0$, when the clock is started.

---

[20] *Science for the Citizen* (Allen and Unwin, London, 1938), p. 747.
[21] "New Vistas for Intelligence" in *Physical Science and Human Values*, ed. E. P. Wigner (Princeton, 1947), p. 144.

[22] In calculus, VELOCITY, v, at an instant, is defined by $v = \frac{ds}{dt}$, and ACCELERATION, a, at an instant, is $\frac{dv}{dt}$ or $\frac{d^2s}{dt^2}$.

In these relations, $\frac{1}{2}a$ is a constant number, since we are assuming a is constant; so, for motion starting from rest,

DISTANCE = (constant number) (TIME)²

or DISTANCE increases in direct proportion to TIME²

or DISTANCE varies directly as TIME²

or DISTANCE ∝ TIME², this being shorthand for any of the versions above.

For example, if a body moving with fixed acceleration falls so far in one second from rest, then it will fall four times as far in two seconds from rest, nine times in three seconds, and so on.

Experimental Investigations

The converse can be shown to be true. IF the distance s varies directly as t^2, THEN the acceleration is constant.[23] That gives us a relation to test in investigating real motions. We can arrange a clock to beat equal intervals of time, and measure the distances travelled from rest by a falling body, in total times with proportions 1:2:3: . . . If the total distances run in the proportions 1:4:9: . . . and so on, we may infer a fixed acceleration. Or, as in one form of laboratory experiment, we can measure the time t for various total distances s, and test the relation s = (constant number) (t^2) by arithmetic, or by graph-plotting.

Over three centuries ago Galileo used this method, though he had neither a modern clock nor graph-plotting analysis. Galileo was one of the first to suggest an accurate pendulum clock, but he probably never made one. All he used to measure time was a large tank of water with a spout from which water ran into a cup. He estimated times by weighing the water that ran out—a crude method yet accurate enough to test his law. However, free fall from reasonable heights takes very little time—the experiment was too difficult with Galileo's apparatus.[24] So he "diluted" gravity by using a ball rolling down a sloping plank. He measured the times taken to roll distances such as 1 foot, 2 feet, etc., from rest.

On the basis of rough experimenting and sturdy guessing, Galileo decided that a ball rolls down a sloping plank with *constant acceleration*. Believing that this would be true for *any* slope, and arguing from one slope to greater slopes and greater still, he expected it to hold for a vertical plank, that is for free fall.[25] The idea of constant acceleration had

been suggested by earlier scientists—who were scorned for it. Galileo did his best to minimize friction, which threatened to complicate matters—though we now know that constant friction would not spoil the simple relationship. His results were rough, but seemed to convince him that his guess was right. It was the simplest kind of accelerated motion he could imagine, and he was probably influenced by the general faith, which has inspired scientists from the Greeks to Einstein, that nature is simple.

Later experiments, with improved apparatus, confirmed Galileo's conclusion: the motion *is* one with constant acceleration, i.e., with $\Delta v / \Delta t$ = constant, in all the following cases:

for a ball or wheel rolling down a straight inclined plank;

for a body sliding down a smooth inclined plank, or a truck with wheels running down it;

for free fall.

Yet each such test has only shown that the acceleration is constant for that one set of apparatus, on that one occasion and within the limits of accuracy of that experiment. If as scientists we want to believe in a general rule inferred from these experiments, if we want to codify nature's behavior in a simple "law" as a starting point for new deductions, then we need a great body of consistent testimony as a foundation for our inference. The more the better, in quantity and variety, and no witness is unwelcome. If any experiment contradicts this general story—and some do—it thereby offers a searching test. "The exception proves the rule" is a fine scientific proverb—though often misunderstood—if "proves" means "tests" (as in "proving-grounds" for artillery, the "proving" of bank accounts). If "proves" had the modern common meaning of "shows it to be right" the proverb would be nonsense.[26] Exceptions do *not* show that the rule is

---

[23] By calculus: if $s = kt^2$, then velocity $\dfrac{ds}{dt} = 2\,kt$;

and acceleration $\dfrac{dv}{dt} = \dfrac{d}{dt}\left(\dfrac{ds}{dt}\right) = 2k$, which is constant.

[24] Galileo's apparatus was rough. He used it to illustrate his argument rather than to measure acceleration.

[25] He convinced himself that the speed acquired by a

body sliding down a frictionless incline depends only on the height, h, not on the length of slope, L. If so, a body falling freely through a vertical height, h, would acquire the same speed, since this would be like a vertical incline. Then he could argue safely from his experiments to vertical fall.

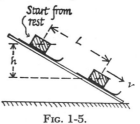

FIG. 1-5.

[26] The original legal meaning is amusing, but irrelevant here: "the quoting of an exception makes it clear that the rule exists."

correct. Exceptions do put a rule to fine tests and show its limitations. They raise the question "What is to blame?" and they lead either to limitations of the rule or to greater care in experimenting. Either way, the rule emerges more clearly established.

The work of many scientists, professional researchers and amateurs such as yourselves, has built up great faith in Galileo's discovery: *bodies falling freely under gravity, and bodies sliding or rolling down a straight slope, under diluted gravity, move with constant acceleration.*

Further experiments show that the acceleration has the same constant value even if the body does not start from rest but is given a push to start it. If it already has speed v_0 when the clock starts, then the simple relation $s = \frac{1}{2} at^2$ no longer holds; we must use $s = v_0 t + \frac{1}{2} at^2$. But the acceleration, a, is the same. It could hardly be different: how could the ball know that it had started with a shove instead of rolling down an earlier piece of the same incline?

The Actual Acceleration

Experiments do more than just assure us the acceleration is constant: they tell us its actual value. If a is constant, then DISTANCE $= (\frac{1}{2}a)$ (TIME)^2, so DISTANCE/TIME$^2 = \frac{1}{2}$ ACCELERATION.

Suppose in some experiment we measured time in seconds and distance in feet, then calculated $\frac{1}{2}a$ and obtained 0.076. Then, in that experiment, $a =$ 0.52 or 2/13. But 2/13 is incomplete—two-thirteenths of what? Such a number is useless unless it carries a tag to show its units. We calculated this number by dividing distance, in feet, by time^2. Since the time is in seconds, the answer must be feet/sec^2. (This is read "feet per second squared" or "feet per second per second.")

Units for Acceleration

Return to the definition of acceleration to look for its units directly;

$$a = \frac{\Delta v, \text{ measured in velocity-units, e.g., feet/second}}{\Delta t, \text{ measured in time-units, e.g., seconds}}$$

$=$ acceleration measured in acceleration-units,

e.g., $\dfrac{\text{ft/sec}}{\text{sec}}$.

Thus we expect to measure acceleration in units such as $\dfrac{\text{ft/sec}}{\text{sec}}$, which we write ft/sec/sec or ft/sec^2.

Scientific Units

In ordinary life, we measure speeds in *feet per second* or in *miles per hour*, and engineers often use these units. We express accelerations in *feet/second per second*, or sometimes in stranger units such as *miles/hour per second*. But scientists all over the world have agreed to use the metric system of units in their measurements, and we shall use one version of this, the Meter-Kilogram-Second system. In this "MKS" system, lengths and distances are measured in meters instead of feet, masses of stuff in kilograms instead of pounds, and times in seconds. A meter is almost 10% longer than a yard, its exact length being defined by a bar of fireproof metal which is carefully preserved, with copies in standardizing laboratories throughout the world. A kilogram is roughly 2.2 pounds, 10% more than 2 pounds.

Acceleration of Free Fall

For free fall, the acceleration can be measured. To show that the acceleration is constant as a body falls faster and faster is difficult, though of course it can be done with modern timing apparatus, some of which can measure to one-millionth of a second. If we *assume* the acceleration is constant, then it is fairly easy to measure its value by timing free fall for one known distance from rest and using the relation $s = \frac{1}{2}at^2$. This leads to $a = 2s/t^2$. As a reminder that we are dealing with a characteristic constant acceleration "due to gravity," we label this particular acceleration "g" and write $g = 2s/t^2$. Using experimental values of s and t we can compute g. However, air friction limits the accuracy; it is difficult to make sure that we start the timing just when the falling body starts from rest, and the time of fall itself is a very short one; so such measurements do not give an accurate value of g. Yet we need to know g accurately for a number of uses in physics. Could we possibly eliminate the effects of friction? And could we lump together many falls, say several thousand, and measure the total time for the whole bunch to obtain the time for one fall with greater accuracy? These look like hopeless ambitions. Yet they can be achieved in a simple, easy experiment which Galileo foreshadowed, and which you will meet.

Measurements give a value about 9.8 meters/sec^2 for g, or 32.2 ft/sec^2. For ordinary calculations, 32 ft/sec^2 will suffice: accurate within 1%.

At the Equator, g is slightly smaller; and at the North Pole g is slightly greater.

Force and Acceleration

We think of a falling body as being pulled down by a force which we call its weight. To hold a body suspended we must support its full weight. If we cut the suspending cord we imagine the weight still acting, now unopposed by our supporting pull. If we suppose the body's weight remains constant while the body is falling, we may picture this constant force "causing" the constant acceleration of free fall. Trucks running down a slope have a smaller acceleration, a fraction of g; but only a fraction of their weight is available to pull them down along the slope. This fraction is, in fact, the hill's "slope" (height/slant length). If you knew this fraction, you could follow Galileo in comparing downhill FORCE and downhill ACCELERATION. What kind of relation would you expect[29] to find between the force and the acceleration? You can see how early experimenters like Galileo could guess at it by studying falling and rolling bodies. That relation, to be discussed soon, is a very important piece of physics, a basic relation governing the motion of stars and the action of atoms, one of obvious importance in engineering.

We will end on a note of doubt. How do you know the weight of a body pulls it while it is falling freely? When you sit on a chair you feel the supporting force of the chair, and you believe you feel your own weight. But if you jump out of a window, do you feel your weight while you are falling? Suppose you jump out of a window with a lump of metal in your hand and try to weigh the lump as you fall. To make the temporary laboratory more comfortable, for a time, suppose you and the lump and the weighing apparatus are enclosed in a vast box which has been dropped from a tower and is falling freely. Suppose the box has no windows. When you release the lead lump inside the box, will it fall to the floor? If you think about this, you will see that gravity will seem to have disappeared. Can you possibly tell whether gravity has really disappeared or whether your laboratory is accelerating downwards? If you cannot tell the difference, *is* there any difference? Discussion of these questions would lead you towards the Theory of Relativity.

---

[29] Do we mean "expect" or "hope"? If *expect*, on what basis? If *hope*, is this scientific or not?

CHAPTER 1 · MANKIND AND THE HEAVENS

"An undevout astronomer is mad."
—EDWARD YOUNG (~1700)

The Beginnings of Man[1]

ASTRONOMY is almost as old as man himself. How old *is* man, and why did he bother about astronomy?

Man began to emerge as a distinct creature several thousand centuries ago. The records of the caves and rocks—the only records stretching so far back—are far from completely explored. Anthropologists warn us not to guess too confidently, and not to guess at all unless we first decide what we mean by Man. What essential characteristics distinguish man from animals? Solving problems? Rats can solve a maze: ants organize war. Using tools? Apes use sticks and stones to solve an immediate problem, and some build simple tree huts. Planning for the future? Perhaps a clear difference begins there. Man makes tools for *future* use. Such planning involves a simple form of reasoning: if . . . then. Man makes arrows for game that *may* come; and he builds graves to comfort the dead *if* they have some future life. Planning for food supply and shelter could lead, with the help of speech, to larger

MAP B. Copernicus lived and worked near Thorn. Kepler was born in Weil and first taught at Gratz.

plans . . . community life . . . traditions . . . laws. . . . Thus, man emerged as a planner and reasoner, a worrier. He used his toolmaking hands and worrying brain in adapting himself to changes of environment. Unlike animals, man changed his living-equipment quickly to meet new conditions: shifts of climate, invasions, floods, famines. Instead of depending on heredity—the chance survival of an animal's mutation in several generations—man controlled his own adaptation. He changed tools, clothing, housing, feeding and defense to meet each new environment. That gave him far greater chances of survival—and hopes of progress.

How old is man? Perhaps as long as 2,000,000 years ago, very primitive tool-making man was showing divergencies from ape-like "cousins." Some 200,000 years ago simple "Neanderthal man" developed along a side-line of our family tree, and was later displaced by our more capable ancestors. He was a simple hunter with rough tools, but he made some use of fire, and he buried his dead with care. There are few signs of our direct ancestors until, say, 100,000 years ago,[2] when men using carefully

MAP. A.

[1] Some of the comments here are drawn from an excellent popular book on primitive man, *Man Makes Himself*, by V. Gordon Childe, published in paperback as a Mentor Book (New York, 1951).

[2] These spans of time may be wrong by a factor of 2. Even if they are true of man's development in one region of the world, they are untrue in others.

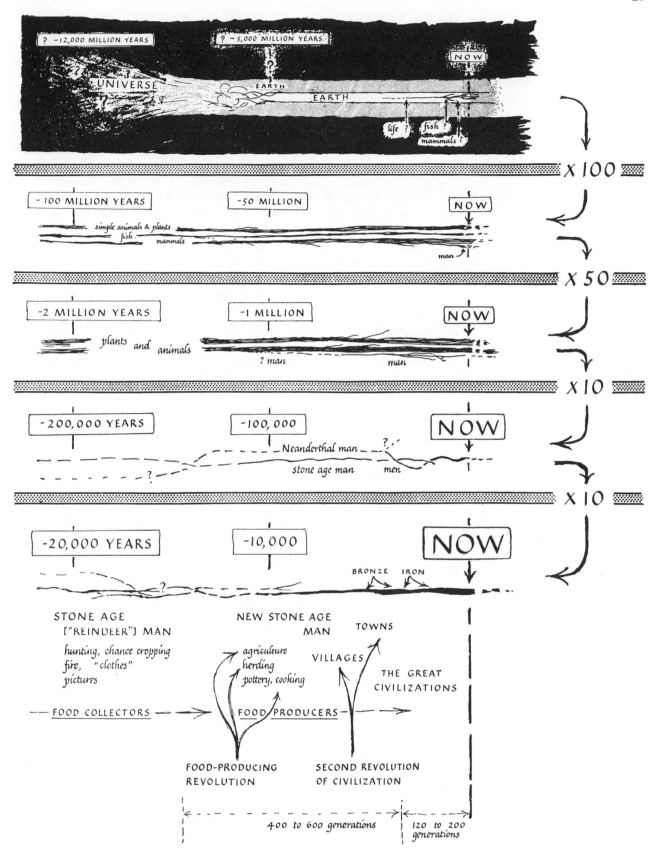

FIG. 12-1. A ROUGH TABLE OF AGES

The times for the stages of man differ greatly with locality. "The age of the Universe" not only seems a fantastic guess but is also entirely dependent on our choice of time-scale: yet astrophysicists are making scientific speculations.

chipped stones pushed their way across Europe. In the next 80,000 years these stone tools and weapons improved, bone needles were made and used, carvings and pictures were added; but man remained a *food-collecting* savage, living in small groups, with leisure when game was plentiful. Stone-age artists made little statues to symbolize fertility and drew animals on cave walls, for simple magic. Some are masterpieces of artistic skill and sympathetic insight.

It was not till some 12,000 years ago that an *age of cultivation* began. Then a new level of human life developed—perhaps even new races of man— in which better tools were used, agriculture began to supplement chance cropping, herding began to replace hunting of wild animals, and pottery and cooking came into use. Village life developed in this *food-producing* culture, and simple trade was carried on.

Then, five or six thousand years ago, a new revolution of man's living started: groups of villages were gathered into states, with farm and town developing their distinct activities—the beginning of the great civilizations. Large cities were built to maintain secondary industries, fed by organized farming outside and enriched by far reaching trade. Craftsmen's knowledge brought metals to replace stone for tools: copper, bronze, then iron. In cities, problems of building and trade and government required arithmetic; geometry; measures of weight, length, area, volume; and timekeeping. Organized agriculture to feed the cities required a good calendar to arrange for planting, animal breeding, and irrigation by river floods. Long trade routes by sea and land required signposts for navigation. Compass, clock, and calendar were essential in the early civilizations, as they are now: *astronomy provided all three.*

Man's Growth

Three thousand years ago, there were flourishing civilizations with good practical knowledge of astronomy. Observations of the Sun, Moon, and stars, recorded, codified, and extended to predictions, provided a daily clock and an accurate calendar of months and years and a compass for steering journeys. *Thirteen thousand* years ago, man had been a simple food-gatherer, using Sun, Moon, and stars as rough guides at most. A great science of astronomy grew up in the ten thousand years between. If that seems long, write it in generations: from savage with simple magic to civilization with working astronomy in 400 generations, another 120

to our knowledge of science today. That is rapid progress, both in power over environment and in intellectual grasp.

The Beginning of Astronomy

Yet, earliest knowledge of the heavens grew slowly. For centuries very primitive man must have watched the stars, perhaps wondering a little, taking the Sun for granted, yet using it as a guide, relying on moonlight for hunting, and even reckoning simple time by moons. Then slowly gathered knowledge was built into tradition with the help of speech. The Sun offered a rough clock by day and the stars by night.[3] For simple geography, sunrise marked a general easterly region, and sunset a westerly one; and the highest Sun (noon) marked an unchanging South all the year round. At night the pole-star marked a constant North.

As the year runs through its seasons the Sun's daily path changes from a low arc in winter to a high one in summer; and the exact direction of sunrise shifts round the horizon. Thus the Sun's path provided a calendar of seasons; and so did the midnight star-pattern, which shifts from night to night through the year.

With the age of cultivation, herding and agriculture made a calendar essential. It was necessary to foretell the seasons so that the ground could be prepared and wheat planted at the right time. Sheep, among the first animals domesticated, are seasonal breeders, so the early herdsmen also needed a calendar. Crude calendar-making seems easy enough to us today, but to simple men with no written records it was a difficult art to be practised by skilled priests. The priest calendar-makers were practical astronomers. They were so important that they were exempt from work with herds or crops and were paid a good share of food, just as in many savage tribes today.

When urban civilizations developed, clear knowledge of the apparent motions of Sun, Moon, and stars was gathered with growing accuracy and recorded. These regular changes were worth codifying for the purpose of making predictions. In the great Nile valley, where one of the early civilizations grew, the river floods at definite seasons, and it was important for agriculture and for safety to predict these floods. And fishermen and other sailors at the mercy of ocean tides took careful note of tide regularities: a shifting schedule of two tides a day, with a cycle of big and little tides that follows the

[3] An experienced camper can tell the time by the stars within ¼ hour.

Moon's month. In cities, too, time was important: clocks and calendars were needed for commerce and travel.[4]

Timekeeping promoted an intellectual development: "In counting the shadow hours and learning to use the star clock, man had begun to use geometry. He had begun to find his bearings in cosmic and terrestrial space."[5]

Astronomy and Religion

Apart from practical uses, why did early man attach such importance to astronomy and build myths and superstitions round the Sun, Moon, and planets?

The blazing Sun assumed an obvious importance as soon as man began to think about his surroundings. It gave light and warmth for man and crops. The Moon, too, gave light for hunters, lovers, travellers, warriors. These great lamps in the sky seemed closely tied to the life of man, so it is not surprising to find them watched and worshipped. The stars were a myriad lesser lights, also a source of wonder. Men imagined that gods or demons moved these lamps, and endowed them with powers of good and evil. We should not condemn such magic as stupid superstition. The Sun *does* bring welcome summer and the Moon *does* give useful light. The simple mind might well reason that Sun and Moon could be persuaded to bring other benefits. The very bright star, Sirius, rose just at dawn at the season of the Nile floods. If the Egyptians reasoned that Sirius caused the floods, it was a forgivable mistake—the confusion of *post hoc* and *propter hoc* that is often made today.

When a few bright stars were found to wander strangely among the rest, these "planets" (literally "wanderers") were watched with anxious interest. At a later time, early civilized man evolved a great neurotic superstition that man's fate and character are controlled by the Sun, Moon, and wandering planets. This superstition of astrology, built on earlier belief in magic, added drive—and profit—to astronomical observation.

Thus the growth of astronomy was interwoven with that of religion—and they still lie very close, since modern astronomy is bounded by the ultimate questions of the beginning of the world in the past and its fate in the future. The next two pages contain speculations on the early stages of that development.[6]

Science, Magic, and Religion

Science began in magic. Early man lived at the mercy of uncontrolled nature. Simple reasoning made him try to persuade and control nature as one would a powerful human neighbor. He tried simple imitative magic, such as jumping and croaking like a frog in the rain to encourage rainfall, or drawing animals on cave walls to promote success in hunting. He buried his dead near the hearth, with logical hope of restoring their warmth; and he gave them tools and provisions for future use. In a way, he was carrying out scientific experiments, with simple reasoning behind them. It was not his fault that he guessed wrong. The modern scientist disowns magicians because they refuse to learn from their results—that is the essential defect of superstition, a continuing blind faith. Primitive man, however, had neither the information nor the clear scientific reasoning to judge his magic.

As he carried out jumping ceremonies, or squatted before magic pictures, man could form the idea of presiding spirits: kindly deities who could help, malicious demons who brought disaster, powerful gods who controlled destiny. Like a child, man tried to please these gods so that they would grant good weather, health, plentiful game, fertility. The original reasons and purposes were then forgotten and the ceremonies continued by habit.

Speech was the essential vehicle of this development. Earliest man, just beginning speech, was forging his own foundations of thought, slowly and uncertainly. Other creatures communicate—bees dance well-coded news of honey, dogs bark with meaning—but man's speech opened up greater intellectual advances. In the course of a long de-

---

[4] The ancient civilizations had no reliable mechanical clocks, but only sand-glasses, simple lamps, and water-clocks. For accurate timekeeping they used the Sun and stars. Pendulum clocks and good portable watches are recent medieval inventions to meet the demands of ocean trade.

[5] Lancelot Hogben, *Science for the Citizen* (Allen and Unwin Ltd., London, 1938). In the early chapters, the author discusses how and why astronomy developed; and he gives a fine account of astronomical measurements and their use in navigation. Some of the material of the present section is drawn from that stimulating book on the social background of scientific discovery.

[6] Such speculative guessing is very risky. It is the misfortune of the young Science of Prehistory that laymen, and even scientists from other fields, think they can guess correctly how man grew and even what he thought. (History suffers similarly from amateur speculation; Education is almost built on it.) Yet here we need some picture of the background of the beginning of science. Make your own speculations, if you prefer; or look at the early chapters of H. G. Wells' *Outline of History*, on which some of the comments here are based. In that book, often criticized for inaccuracy of fact and view, the general reader finds what the experts fail to give him, a connected story—though a risky one.

velopment, it not only gave him a rich vocabulary for communicating information but it enabled him to store information in tales for later generations; and then it blossomed into a higher intellectual level with *words for abstract ideas.* Thus speech opened up a new field of *ideas and reasoning.* Of course, this development did not happen suddenly.[7] Early verbal thinking must have been crude and confused, with reasoning left unfinished, and words mistaken for the things they represent—as they are in children's thinking today and even in simple people's attitude to science.

With speech could come the beginnings of religion and science: rules of conduct for the individual and the community; and, in a different sense, rules for nature. Even before speech developed, family life involved obedience; but speech could hand on the tradition of "the strong father, the old man whose word was law, whose possessions must not be touched," and of a kindlier mother-figure. As families gathered into groups . . . villages . . . tribes . . . , these traditions crystallized into law and custom that restricted the individual's freedom for the general good. Out of such feelings and traditions, out of hopes of success and fears of disaster, grew a sense of being bound together in a community, a sense of *religion.*

Primitive religion wove together myths of gods, magic ceremonies, and tales of nature, in attempts to codify both the natural world and man's growing social system. Astronomy played an important part in this ceremonial religion. The priests—wise men of the village or tribe—were the calendar-makers, the first professional astronomers. Their successors were the powerful priesthood of the first urban civilizations. In early Babylon, for example, the priests were bankers, physicians, scientists, and rulers—they *were* the government. In their knowledge, and that of their craftsmen, there were the foundations of many sciences. The practical *information* was there, at first unrecognized, then held secret, then published in texts. Was it *science?*

[7] Anthropologists warn us: do not assume that primitive savage communities today give a reliable picture of our equally primitive ancestors long ago. Contemporary savages may be primitive in technology and yet maintain a complex of customs or religion developed over thousands of years. It may be safer to base surmises on the behavior of civilized children.

Science, the Art of Understanding Nature

Curiosity and collecting knowledge go back before earliest man. Primitive man collected knowledge and used it: a beginning of applied science. Then, as a reasoner, he began to organize knowledge for use and thought. That is a difficult step, from individual examples to generalization—watch a child trying to do it. It is difficult to grasp the idea of a common behavior or a general law or an abstract quality. Yet that is the essential step in turning a "stamp collection" of facts into a piece of science. Science, as we think of it and use it now, never was just a pile of information. Scientists themselves, beginning perhaps with the early priesthood, are not just collectors. They dig under the facts for a more general understanding: they extract general ideas from observed events.

Scientists feel driven to *know*, know what happens, know how things happen; and, for ages, they have speculated why things happen. That drive to know was essential to the survival of man—a generation of children that did not want to find out, did not want to understand, would barely survive. That drive may have begun with necessity and fear; it may have been fostered by anxiety to replace capricious demons by a trustworthy rule. Yet there was also an element of wonder: an *intellectual delight* in nature, a delight in one's own sense of understanding, a delight in creating science. These delights may go back to primitive man's tales to his children, tales about the world and its nature, tales of the gods. We can read wonder and delight in stone-age man's drawings; he watched animals with intense appreciation and delighted in his art. And we meet wonder and delight in scientists of every age who make their science *an art of understanding nature.*

As scientists, we have travelled a long way from capricious gods to orderly rules; but all the way we have been driven by strong forces: a sense of urgent curiosity and a sense of delight.

Fear and anxiety, wonder and delight—these are two aspects of *awe,* mainspring of both science and religion. Lucretius held, 2000 years ago, that "Science liberates man from the terror of the gods." Walt Whitman grieved for man's anxiety while he rejoiced in the scientist's delight:

I believe a leaf of grass is no less than the journeywork of the stars,
And the pismire is equally perfect, and a grain of sand, and the egg of the wren,
And the tree-toad is a chef-d'œuvre for the highest,
And the running blackberry would adorn the parlors of heaven,
And the narrowest hinge in my hand puts to scorn all machinery,
And the cow crunching with depressed head surpasses any statue,
And a mouse is miracle enough to stagger sextillions of infidels,
And I could come every afternoon of my life to look at the farmer's girl boiling
    her iron tea-kettle and baking shortcake.

.   .   .   .

I think I could turn and live awhile with the animals . . . they are so placid
    and self-contained,
I stand and look at them sometimes half the day long.
They do not sweat and whine about their condition,
They do not lie awake in the dark and weep for their sins,
They do not make me sick discussing their duty to God,
Not one is dissatisfied . . . not one is demented with the mania of owning things,
Not one kneels to another nor to his kind that lived thousands of years ago,
Not one is respectable or industrious over the whole earth.

Leaves of Grass
Doubleday, N.Y.

~~~~~~

"To speculate without facts is to attempt to enter a house of which one has not the key, by wandering aimlessly round and round, searching the walls and now and then peeping through the windows. Facts are the key."   —JULIAN HUXLEY, *Essays in Popular Science*

~~~~~~

Facts

Before showing how astronomy was organized into great schemes of thought, we shall review the knowledge that you—or primitive man—could gain by watching the heavens. If you have lived in the country you will be familiar with most of this. If you have grown up in a town this will seem a confusing pile of information unless you go out and watch the sky—now is the time to observe.

The Sun as a Marker

Each day the Sun rises from the eastern horizon, sweeps up in an arc to a highest point at noon, due South,[1] and down to set on the western horizon. It is too bright for accurate watching, but it casts a clear shadow of a vertical post. At noon, mid-day between sunrise and sunset, the shadow is shortest and points in the same direction, due North, every day of the year. The positions of the noon Sun in the sky from day to day mark a vertical "meridian" (= mid-day) plane running N-S.

In winter the shadows are longer because the Sun sweeps in a low arc, rising south of East and setting south of West.[1] In summer the arc is much higher, shadows are shorter, daylight lasts longer. Half-way between these extremes are the "equinox" seasons, when day and night are equally long and the Sun rises due East, sets due West.

On the horizon—simple man's extension of the flat land he lived on—sunrise marked a general eastern region for him and the exact place of sunrise showed the season. The length of noonday shadows also provided a calendar. The shadow of a post made a rough clock. Though it told noon correctly, its other hours changed with the seasons—until some genius thought of tilting the post at the latitude-angle (parallel to the Earth's axis) to make a true sundial.

Stars

The stars at night present a constant pattern, in which early civilizations gave fanciful names to prominent groups (constellations). The whole pat-

[1] This description applies to the northern latitudes of the early civilizations.

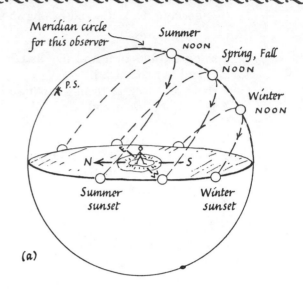

(a)

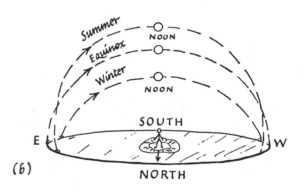

(b)

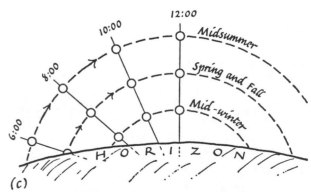

(c)

FIG. 13-1. THE PATH OF THE SUN IN THE SKY CHANGES WITH THE SEASONS

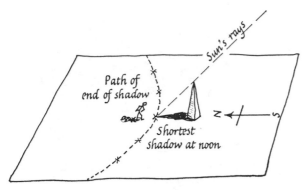

FIG. 13-2. NOON. Sunlight casts the shadow of a pillar on horizontal ground. The shadow is shortest at noon.

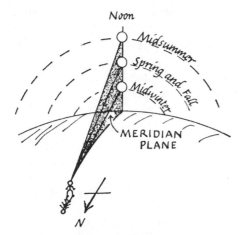

FIG. 13-3. MERIDIAN. Noon-day Sun is due South (or due North). The meridian (= mid-day) plane is a vertical plane passing through the Sun's noon positions.

FIG. 13-5. A PHOTO OF THE SKY NEAR POLE STAR
Taken with an eight-hour exposure. The Pole Star itself made the very heavy trail near the center.
Photo from Lick Observatory.

tern whirls steadily across the sky each night, as if carried by a rigid frame. One star, the pole-star, stays practically still while the rest of the pattern swings round it. Watch the stars for a few hours and you will see the pole-star, due North, staying still while the others move in circles round it. Or, point a camera at the sky with open lens, and let it take a picture of those circles. Night after night, year after year, the pattern rotates without noticeable change. These are the "fixed stars."[2] The pole-star is due North, in the N-S meridian plane of the noonday Sun. The star pattern revolves round it at an absolutely uniform rate. This motion of the stars gave early man a clock, and the pole-star was a clear North-pointing guide.[3]

The simplest "explanation" or descriptive scheme for the stars is that they are shining lights embedded in a great spinning bowl, and we are inside at the center. That occurred to man long ago, and you would feel it true if you watched the sky for many nights. It was a clever thinker who extended the bowl to a complete sphere, of which we see only half at a time. This is the celestial sphere, with its axis running through the pole-star and a celestial equator that is an extension of the Earth's equator.

[2] However, if you could live for many centuries you would notice changes in the shape of some constellations. Stars are moving.
[3] The procession of the equinoxes carries the Earth's spin-axis-direction slowly round in a cone among the stars, so only in some ages has it pointed to a bright star as pole-star. (See sketches in later chapters.) It is near to one now, our present pole-star, and was near to another when the pyramids were built. In A.D. 1000 there was no bright pole-star—perhaps that lack delayed the growth of navigation.

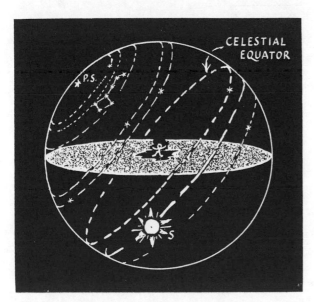

FIG. 13-4. THE STAR PATTERN REVOLVES

The celestial sphere revolves steadily, carrying all the stars, once in 24 hours. The Sun is too dazzling for us to see the stars by day, so we only see the stars that are in the celestial hemisphere above us at night, when the Sun is in the other hemisphere below. The Sun's daily path across the sky is near the celestial equator; but it wobbles above and below in the course of the year, 23½° N in summer, 23½° S in winter.

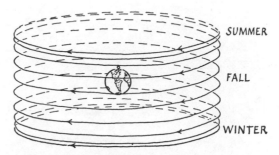

FIG. 13-6c. THE SUN'S SPIRAL OF CIRCLES, in course of half a year of seasons.

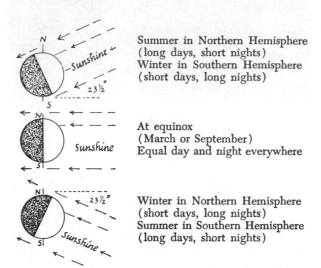

Summer in Northern Hemisphere (long days, short nights)
Winter in Southern Hemisphere (short days, long nights)

At equinox (March or September) Equal day and night everywhere

Winter in Northern Hemisphere (short days, long nights)
Summer in Southern Hemisphere (long days, short nights)

FIG. 13-6a. EARTH AND SUNSHINE
Day and night at various seasons

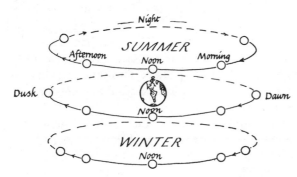

FIG. 13-6b. PATH OF SUN.
Viewed from stationary Earth, at various seasons. The Sun-positions are labeled noon, afternoon, etc., for observers in the longitude of New York. If such an observer could watch continuously, *unobstructed by the Earth*, he would see the Sun perform the "spiral of circles" sketched in FIG. 13-6c, during the course of 6 months from summer to winter; then he would see the Sun spiral upwards, revolving the same way, from winter to summer.

Though the pattern of stars has unchanging shape, we do not see it in the same position night after night. As the seasons run, the part of the pattern overhead at midnight shifts westward and a new part takes its place, a whole cycle taking a year. Stars that set an hour after sunset are 1° lower in

the West next night and set a few minutes earlier; and two weeks later they are level with the Sun and set at sunset. Thus, in 24 hours the celestial sphere makes a little more than one revolution: 360° + about 1°. It is moving a little faster than the Sun, which makes one revolution, from due South to due South in the 24 hours from noon to noon. The celestial sphere of stars makes one extra complete revolution in a year.

Sun and Stars

This difference between the Sun's daily motion and that of the stars (really due to the Earth's mo-

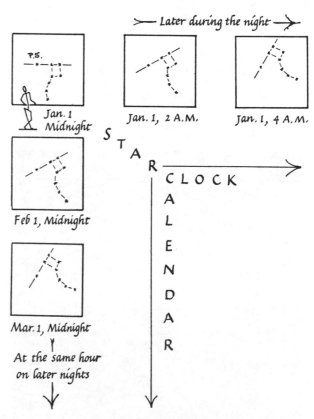

FIG. 13-7. THE STAR-PATTERNS
keep constant shape but revolve nightly and advance 30° per month, compared with Sun's noon and midnight.

tion in its orbit round the Sun) was obvious, and suggested that the Sun is moved by a separate agent. The Sun-god became a central figure in many primitive religions, and his travels were carefully traced by shadows and recorded by alignments of great stones in ceremonial temples.

Instead of saying the star pattern "gains" (like a clock running fast) 1° a day, we take the very constant star motion as our standard and say the Sun lags behind it 1° a day. We may stick the Sun as well as the stars on the inside of the celestial sphere; but since the Sun lags behind the stars it does not stay in a fixed place in the starry sphere; it crawls slowly *backwards* round the inside of the sphere, making a complete circuit in a year. Thus we can picture the Sun's motion compounded of a *daily motion shared with the stars* of the celestial sphere and a *yearly motion backward through the star pattern*.

Ecliptic and Zodiac

This is a sophisticated idea, a piece of scientific analysis: to separate out the Sun's yearly motion from its daily motion across the sky with the star-pattern. Once this idea was clear it was easy to map the Sun's yearly track among the stars—not directly, because the Sun outglares the nearby stars by day, but by simple reference to the pattern of stars in the sky at midnight instead of noon. The Sun's yearly track is not the celestial equator but a tilted circle making 23½° with the equator. It is this tilt that makes the Sun's *daily* path across the sky change with the seasons. At the equinoxes the Sun's

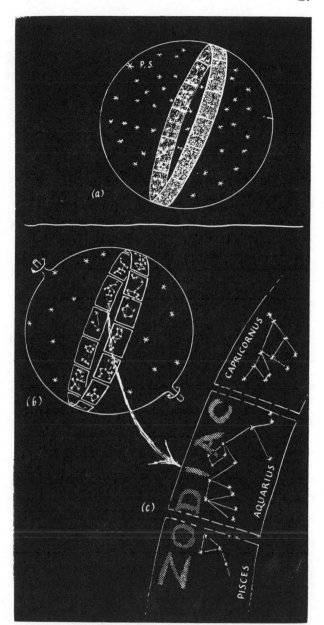

FIG. 13-9. THE ZODIAC, a belt of the celestial sphere, tilted 23½° from the equator. The Sun's yearly path (the ecliptic) runs along the middle line of this belt. The paths of Moon and planets lie within this belt. The Zodiac was divided into twelve sections named after prominent star-groups or constellations.
(Zodiac patterns after H. A. Rey, *The Stars.*)

yearly track crosses the equator. In summer, the tilted track has carried the daily path 23½° higher in the sky and in winter 23½° lower. This tilted yearly track is called the *ecliptic*.

As the Sun travels round the ecliptic in the course of the year, it passes through the same constellations of stars at the same season year after year. This broad belt of constellations containing the ecliptic is called the *zodiac*. The constellations in it were given special names long ago by the astrologer priests, a named group for each month in the year.

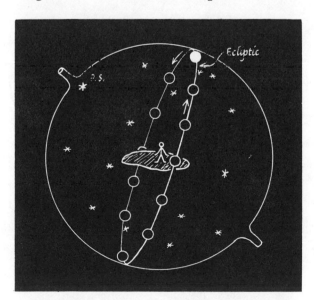

FIG. 13-8. THE ECLIPTIC, the Sun's track through the star pattern in the course of a year. Here the daily motion is imagined "frozen."

Moon

The Moon is obviously moving round the Earth and illuminated by sunlight. Watch it for a week or two; think where the Sun is each time and see if it accounts for the Moon's lighting. The Moon

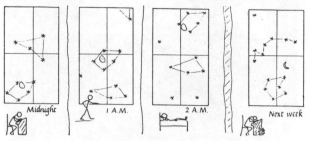

FIG. 13-10. THE MOON'S MOTION
The Moon, while moving across the sky with the stars each night, slips rapidly backward through the star-pattern, a whole circle in a month.

swings across the sky with the neighboring stars, but even in one night it lags noticeably behind the stars. It lags much quicker than the Sun, 90° in a week; right round the star-pattern in a month. Its monthly track is tilted about 5° from the ecliptic, but it is still within the broad band of the zodiac.

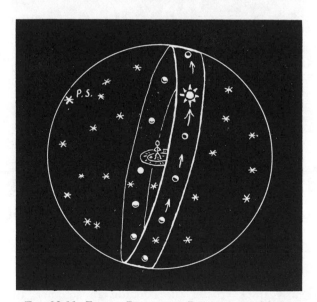

FIG. 13-11. ZODIAC BELT WITH POSITIONS OF MOON, IN VARIOUS PHASES, IN THE COURSE OF A MONTH
The daily motion of the celestial sphere is "frozen" here.

Eclipses

Eclipses are impressive events. Something seems to take a bite out of the Sun or Moon. A total eclipse of the Sun is awe-inspiring even to educated people—daylight disappears and it grows strangely cold.

"In the priestly calendar lore, magic and genuine science were inextricably entangled. . . . As liaison officers to the celestial beings, the priests found it paid to encourage the belief that nature can be bought off with bribes like a big chief. One of their most powerful weapons was their ability to forecast eclipses. Eclipses were indisputable signs of divine disapproval, and divine disapproval provided a cogent justification for raising the divine income-tax. No practical utility other than the advancement of the priestly prestige and the wealth of the priesthood can account for the astonishingly painstaking attention paid to these phenomena."[4]

Later, men realized that eclipses are just shadows. When the Moon is eclipsed the Sun throws the Earth's shadow on the Moon. The Sun is eclipsed when the Moon *gets in the light* between us and

FIG. 13-12. ECLIPSES
NOT TO SCALE. See FIG. 14-19b, c.

(a)

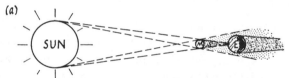

FIG. 13-12a. Eclipses of the Sun occur when the Moon gets in the light between Sun and Earth. The Moon's size and distance are such that total eclipses are just possible.

(b)

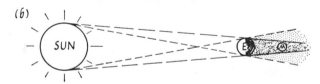

FIG. 13-12b. Eclipses of the Moon occur when the Moon passes through the Earth's shadow.

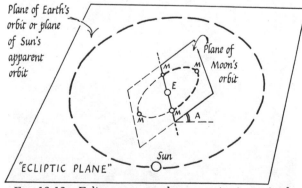

FIG. 13-12c. Eclipses occur only at certain times. Angle A is about 5°. However, as the line where the Moon's orbit-plane cuts the ecliptic plane slowly moves around —due to perturbing attractions—eclipses do not always occur at the same season of our year.

[4] Lancelot Hogben, *Science for the Citizen*, Allen and Unwin, Ltd., pp. 43-44.

the Sun, and we are in the Moon's shadow, which sweeps rapidly across (part of) the Earth.

For an eclipse, Sun, Moon, and Earth must be in line. There is a chance of this only when the Moon in its tilted orbit crosses the ecliptic plane, which by definition contains Sun and Earth—hence the name "ecliptic." Even then, the necessary alignment is rare. An eclipse of the Moon is a shadow *on the Moon*, and looks the same from everywhere on the Earth. So an eclipse of the Moon observed from different parts of the Earth occurs at different times *by the local clocks*—a proof that the Earth is round, not flat.

Calendar Periods

Day. The Sun's motion from noon to noon marks an almost constant day. However, it varies slightly; sundial noon runs ahead of a constant clock at some seasons and behind at others, sometimes by many minutes. The Sun's apparent motion along the ecliptic is not exactly regular through the year—it moves faster in winter—so its daily motion shows slight changes. (The Moon's motion shows even more complex irregularities.) The stars' motion round the celestial axis through the pole-star defines a constant, slightly shorter, day—man's ultimate standard of timekeeping until the perfection of electronic clocks.

Month. The Moon was probably the first calendar-marker for men. The month from full moon to full moon is about 29½ days. The full moon is exactly opposite the Sun, so that is the month judged relative to the Sun. In 29½ days the Sun travels almost 29° along the ecliptic, so the Moon makes more than one revolution, 360° + 29°, relative to the stars to catch up with the Sun. Relative to the stars as fixed markers, the Moon takes 27.3 days for one revolution. We use a 29½-day month, like the early-calendar makers, to predict full moon, new moon, etc.; but we shall use the 27.3-day period when we calculate the Moon's motion under gravity.

Year. The idea of a year grew up as:
(a) the repeat-time of the seasons
(b) the time the Sun takes to return to the same place among the stars (or the stars to return to the same midnight position in the sky). This differs slightly from (a).
(c) a period of 12 (or 13) moon-months. Easy to observe, such a year soon gets out of step with the solar year of seasons.

Planets

A few bright "stars" do change their positions, and move so unevenly compared with Sun, Moon, and the rest that they are called *planets*, meaning wanderers. These planets, which look like very bright stars, with less twinkling, wander across the sky in tracks of their own *near the ecliptic*. They

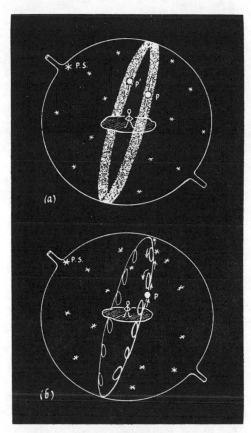

FIG. 13-13. THE PATH OF A PLANET
All the planets wander through the star pattern in a belt near the ecliptic—the zodiac belt.
(a) General region of a planet's path—the zodiac belt.
(b) In detail, a planet's path has loops—an epicycloid seen almost edge-on.

follow the general backward movement of the Sun and Moon through the constellations of the zodiac, but at different speeds and with occasional reverse motions. Primitive man must have observed the brighter planets but cannot have got any good use from his observations, unless like eclipses they were used to impress people.

Zodiac

Thus the zodiac belt includes the Sun's yearly path, the Moon's monthly path, and the wandering paths of all the planets. In modern terms, the orbits of Earth, Moon, and planets all lie near to the same plane. Astrology assigned fate and character by the

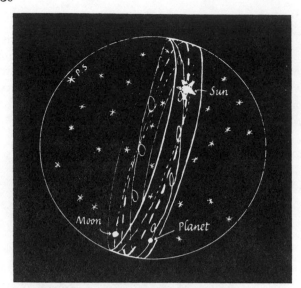

FIG. 13-14. ZODIAC BELT WITH PATHS OF SUN (in one year), MOON (in one month), and a specimen PLANET (in planet's "year"). The daily motion of the celestial sphere is "frozen" here.

places in the zodiac occupied by Sun, Moon, and planets at the time of a man's birth.

The Planets and Their Motions

Five wandering planets were known to the early civilizations, in addition to the Sun and Moon which were counted with them. These were:

Mercury and Venus, bright "stars" which never wander far from the Sun, but move to-and-fro in front of it or behind, so that they are seen only near dawn or sunset. Mercury is small and keeps very close to the Sun, so it is hard to locate. Venus is a great bright lamp in the evening or morning sky. It is called both the "evening star" and the "morning star"—the earliest astronomers did not realize that the two are the same.

Mars, a reddish "star" wandering in a looped track round the zodiac, taking about two years for a complete trip.

Jupiter, a very bright "star" wandering slowly round the ecliptic in a dozen years.

Saturn, a bright "star" wandering slowly round the ecliptic, in about thirty years.

Jupiter and Saturn make many loops in their track, almost one loop in each of our years.

When one of the *outer* planets, Mars, Jupiter or Saturn, makes a loop in its path it crawls slower and slower eastward among the stars, comes to a stop, crawls in reverse direction westward for a

while, comes to a stop, then crawls eastward again, like the Sun and Moon.[5]

The sketches show the looped tracks of planets through the star patterns. Once noticed, planets presented an exciting problem to early scientists: what gives them this extraordinary motion? From now our main concern in this section on astronomy will be: how can we explain (produce, predict) the strange motions of the planets, which excited so much wonder and superstition? We study this to show how scientific theory is made.

Epicycloid

Nowadays we call the looped pattern of a planet's track an epicycloid (from the Greek for outer-circle) because we can imitate it by rolling a little circle round the circumference of a big one. Fig. 13-17 shows a scheme to manufacture an epicycloid which imitates a planet's track. A large wheel W spins steadily round a fixed axle. At some point A on its rim, there is an axle carrying a small wheel,

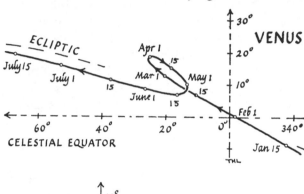

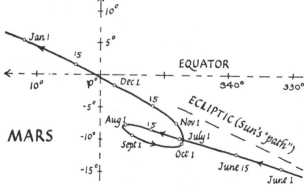

FIG. 13-15. PATHS OF PLANETS THROUGH THE STAR PATTERN
(a) Venus (January-July 1953)
(b) Mars (June-December 1956)
(The ecliptic is the Sun's apparent path. The planets' orbits run close to the ecliptic, because the planes of those orbits are close to the plane of the Earth's orbit (or the Sun's apparent orbit, the ecliptic).

---

[5] Remember that though the Sun sweeps from East to West in its daily motion with the stars, it crawls backward, or West to East, in its yearly motion round the ecliptic. So do planets.

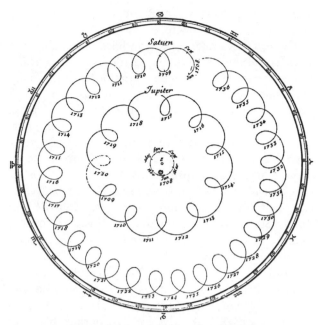

FIG. 13-16. PATHS OF PLANETS IN THE SKY.
This sketch shows the apparent paths of Jupiter and
Saturn, plotted for many years, as they would appear to
an observer attached to the Earth but viewing them from
far out from the Earth, so that the epicycles are seen
face-on, without the foreshortening really observed.
The apparent orbit of the Sun is also shown.
The Earth is at the center. When the astronomer Cassini
constructed this diagram in 1709 he used Copernican
measurements of orbit sizes.

FIG. 13-17. MACHINE FOR DRAWING EPICYCLOIDS

w, which spins steadily, much faster than W. Then
a point P on the rim of the small wheel traces out
an epicycloid. The planetary path we observe is
like this epicycloid seen obliquely, as if the whole
contraption of wheels were at eye level. (There is
a strong hint in this model that a planet's
apparent motion is compounded of two circular
motions. The hint grows stronger when we find one
of those motions is a yearly one for every planet,
taking almost one of our years. The ancient astronomers did not take this hint.)

Observation

Many city-dwellers today take little notice of the
sky, but to anyone living out of doors at night the
planets are obvious strange bright things. Once you
have seen them, you are not likely to miss them
again. With even the simplest telescope or field

glasses you can see surprising details: the crescent-
shaped phases of Venus, Jupiter's moons, perhaps
Saturn's rings. With a telescope the planets look
bigger, while the fixed stars do not. This is because
the planets are much nearer. The fixed stars are
farther away and bigger, but so much farther than
bigger that they look like points.[6]

Planets and Stars

We now know that these nearby wanderers, the
planets, are things of much the same size as our
Earth, and, like the Earth and the Moon, shine only
by reflected sunlight. (As a test of this we now
examine their light with a spectroscope and find it
carries all the characteristic absorption-lines of sunlight.) The fixed stars, however, are lights in their
own right. They are white-hot furnaces like our
Sun (and the spectroscope tells us how they differ
in composition and temperature).

Parallax

You and I know the Earth moves round the Sun,
swinging 186,000,000 miles across its orbit in six
months. Then we ought to see some changes in the
pattern of stars when we make that huge move—
parallax shifts as they are called. Try this *personal
experiment on parallax.* Look at a group of people,
or chairs or other objects at different distances. Walk
to-and-fro sideways, or walk round in a circular
orbit, and notice how the relative positions change
in the group. Those in the remote background, seem
to stay still, while nearer ones move in little orbits
against the background, with a motion the reverse
of yours. Such *parallax* shifts are used unconsciously
by people in judging distances by a wagging head;
and modern astronomers use them to judge the distance of the Moon, planets, and stars.

Even if the stars were all embedded in a single
spherical bowl, the move across the Earth's orbit
would bring us nearer some of the bowl and distort
the pattern by foreshortening. The ancient astronomers saw no such changes and concluded that the
Earth must be at rest at the center of the universe.
The only alternative explanation was that the stars
are infinitely far away compared with the diameter
of the Earth's orbit. Nowadays very delicate telescopic measurements show that there *are* small
parallax shifts, which place even the nearest stars

[6] Telescopes fail to show the real *size* of stars. Even the
most powerful telescope would still show stars as points but
for the spreading of light waves, which affects the images
formed by all optical instruments and makes a small disc
pattern for each star—the *bigger* the telescope, the *smaller*
the disc.

at vast distances. Much easier measurements place the planets a million times closer. If we could measure distances by clocking a flash of light from each body to us, we should find that light takes 8 minutes from the Sun, a few minutes from the nearer planets and a few hours from the farthest, while from the nearest fixed star it takes several years.

Early Progress

Early astronomy, then, had three driving motives:
(a) practical uses: for compass, clock, and calendar
(b) magic to impress people; and astrology to predict fates and good and bad luck—such beliefs made many a king in later ages support a good astronomer
(c) pure scientific interest: as men grew up, there were scientists like those today whose delight in nature and interest in understanding are the driving forces.

Astronomy in the Early Civilizations

[It is difficult to tell who made the great discoveries because these were probably made in stages and then spread slowly, being re-discovered and claimed many times. Therefore, the notes which follow are not reliable and are offered only as general indications. They cover the development of astronomy from the age when it mattered for corn and animals to the stage when it had taken its place as a science. Here and later, we give short notes instead of a developed account.]

Urban civilizations developed in several great river valleys 5000 or more years ago. Much "applied science" had already been discovered—a few thousand years before then—such as: artificial irrigation of crops by canals and ditches; plow, sailboat, wheeled vehicles; use of animals for motive power; production and use of copper, bricks, glazes; and, finally, a solar calendar, writing, a number system, and the use of bronze.[7]

Sumerians, Babylonians, and Chaldeans (Three separate peoples in Mesopotamia)

By 4000 years ago (2000 B.C.) there were flourishing towns with extensive trade. They had excellent commercial arithmetic that was almost algebra: they could solve problems leading to quadratics, even cubics; they knew value of $\sqrt{2}$ accurately, but took π as roughly $= 3$; and they used similar triangles and knew Pythagoras' rule. They had good weights and measures, sundials, and water-clocks. Records

[7] See V. Gordon Childe, *Man Makes Himself* (Mentor Books, New York, 1951).

and teaching texts were stamped in clay tablets, which we can now decipher.

Their astronomical observations were not the wonders often claimed, but made a good working basis for calendars. Near the equator the Sun's daily path does not provide an obvious calendar, as it does farther North. The Moon is much easier. So the Babylonians based their calendar on new Moons; but they had to reduce that to a solar calendar of seasons for agriculture and seasonal ceremonies. This required careful observations of Sun and Moon. Positions were mapped with reference to the zodiac, which was divided into 12 sections. Stars were catalogued, eclipses recorded, planets observed, and the returns of the planet Venus specially studied.

A thousand years later, the Babylonians developed a marvelous mathematical system for predicting the motions of Sun and Moon accurately. This consisted, essentially, of rules for calculating, like zig-zag time-graphs of the uneven motions. These were empirical, with no theoretical basis, but they maintained an accurate calendar and could even predict probable eclipses. A similar scheme, with rougher interpolation, gave positions of planets. Belief in omens (prophetic signs) flourished, and astrology took a strong hold.

Egyptians

More than 4000 years ago, the Egyptians were living a comfortable life—with more friendly gods—as the Nile floods renewed the soil's fertility every year. Their mathematicians served magic and commerce. Their papyrus texts dealt with measuring corn stores, dividing property, building an exact pyramid. Their magnificent building projects necessitated good mathematics for the organization of the work and the management of armies of workmen. They had good weights and measures and ingenious water-clocks.

Their astronomy was simpler than the Babylonians'. They had an efficient solar year of 12 months of 30 days $+$ 5 extra days; so they paid less attention to eclipses, the Moon, and planets. The Sun-god was supreme in the state religion. Two thousand years later they recorded accurate planet observations, probably for astrology.

Greeks

Some 3000 years ago, Greek civilization began to evolve. It produced mathematicians, scientists, philosophers, who made such important advances that we shall spend a chapter on them—even so, the choice of the few names to be mentioned is unfair.

PROBLEMS FOR CHAPTER 2

1. During the winter, in the Northern Hemisphere, the Earth is actually a little *nearer* the Sun than in the summer. Why then is the winter colder?

2. APPARENT MOTIONS

Contrast prevents us seeing the stars in daytime. Suppose we could see them in daytime: then we should see some pattern of stars near the Sun.

(a) Suppose we could note that pattern at noon in June. When should we see the same star pattern in the same position in the *midnight* sky?
(b) What path should we see the Sun take relative to the unchanging patterns of stars, from month to month (disregarding *daily* motion of stars, etc.)?
(c) Describe the path of an "outer" planet such as Jupiter or Mars relative to the unchanging patterns of stars (disregarding *daily* motion of stars, etc.)?
(d) Describe the path of an "inner" planet such as Venus, relative to the stars.

3. FINDING LATITUDE AND LONGITUDE

(a) State rough values for the latitude of: New York, San Francisco, London, North Pole, Arctic Circle, equator.
(b) State rough values for the longitude of New York, San Francisco, London, Tokyo.

(c) Suppose you are making an exploration in a small boat and are wrecked on an unknown desert island, far off your course. You wish to find your position, but have no radio or other modern electronic equipment, and no special instrument such as a sextant. You do have a simple stick with markers for sighting stars etc. and a plumb line and a protractor for measuring angles. Explain how you would estimate the following: (Say what measurements you would make and how you would treat them. Give a practical explanation that an untrained sailor could use—avoid technical phrases such as "obtain a fix").
(i) your latitude by observing star(s) on a clear night
(ii) your latitude by observing the Sun
(iii) your longitude by observing Sun or stars. (For this a certain auxiliary instrument is *essential*. What instrument?)

(d) (i) How could accurate *predictions* of eclipses of the Moon help in a rough determination of longitude in remote places?
(ii) Why was this use of eclipses seriously considered in ancient times?

How great love is, presence best tryall makes,
But absence tryes how long this love will bee;
To take a latitude,
Sun, or starres, are fitliest view'd
At their brightest, but to conclude
Of longitudes, what other way have wee,
But to marke when, and where the darke
eclipses bee?

—John Donne (∼ 1600)

CHAPTER 3 · GREEK ASTRONOMY:
GREAT THEORIES AND GREAT OBSERVATIONS

"If science is more than an accumulation of facts; if it is not simply positive knowledge, but systematized positive knowledge; if it is not simply unguided analysis and haphazard empiricism, but synthesis; if it is not simply a passive recording, but constructive activity; then, undoubtedly [ancient Greece] was its cradle."

—GEORGE SARTON*

Theory, a House for the Facts, "To Save the Phenomena"

Astronomical knowledge grew up with the early civilizations from simple noticing to systematic observing that provided an official priesthood of astronomers with material for calendar-making on one hand, and on the other a growing tangle of superstitious astrology. With this knowledge came stories to teach children or reassure simple folk. Describing the Sun as a god, worshipping the planet Venus, telling of the "abode of the blessed" above a crystal globe of stars: these were not merely superstitious myths, they were the forerunners, too, of theoretical science. They were not real science: their relationship with fact was thin and fanciful; but they set the pattern of a speculative scheme to "explain" the facts. When Greek civilization formed out of neighboring groups, the wisest thinkers brought a new attitude to science: they sought *general* schemes of explanation that would appeal to the inquiring mind, not simple myths to satisfy public curiosity. Their aim, as they put it, was to "save the phenomena," or save the appearances, meaning to make a scheme that would *account for the facts*. This was a grander business than either collecting facts or telling a new tale for each fact. This was an intellectual advance, the beginning of great scientific theory.

The earliest of the Greek "natural philosophers" gave simple pictures of the structure of the Universe, but as more information was gathered and intellectual tradition grew they evolved schemes to save the phenomena in detail: first simple tales to tell about the Earth, then fuller ones to explain the motion of the sky as a whole and the detailed motions of the Sun, Moon and planets.

At each stage, these philosophers tried to start

Introduction to the History of Science (1927), Vol. 1, page 8, Carnegie Institute of Washington.

with a few simple assumptions or general principles and draw from them as logically as possible a complete "explanation"—or setting forth—of the observed behavior. This explanation would serve to coordinate the information and to make future predictions, but above all to give a feeling that there is a *pattern* that holds diverse behavior together, that *nature makes sense*. Although some of the search for a good scheme was prompted by practical needs such as calendar-making, this delight in a unified clear explanation went far beyond that. Driven by an urge to ask WHY, the Greek philosophers were seeking and making scientific theory. Though our modern tradition of experimenting and our modern wealth of scientific tools have made great changes, we still hold the Greek delight in a theory that will save the phenomena.

This chapter gives an account of some of the Greek scientists. Watch how they built their theory.

Early Greek Astronomy

As Greek civilization developed, some 3000 years ago, the poets (Homer) told the history of earlier neighbors and tried to answer some of the great WHY questions that intelligent people were asking about mankind and about the world. The Earth was pictured as an island surrounded by a great river and covered by a huge bell of the heavens. The home of the gods was at the "ends of the Earth." Hell or the land of the dead was also at the ends of the Earth, or perhaps underneath. A daily Sun rose from the surrounding river and swept over the vault above.

By about 2500 years ago, we hear of great "natural philosophers" telling fuller stories with clearer thinking.

Thales ($\sim$ 600 B.C.) was a founder of Greek science and philosophy. In later centuries his reputation as one of the "seven wise men" grew so fabulous that impossible feats were ascribed to him, such as predicting a solar eclipse. He collected geometrical knowledge, perhaps from Egypt, and began to reduce geometry to a system of principles and deduction—the beginning of a science that Euclid was to bring to full flower. He thought the Earth was a flat disc floating on water; yet he knew the Moon shines by reflected sunlight, so he had applied reasoning to common observations. He is said to have known that lodestone, native rock magnet, will attract pieces of iron; and he is rumored to have discovered electricity by rubbing amber (= "electron" in Greek). Moreover, he went beyond these bits of knowledge and set forth a general explanation of the Universe: that water is

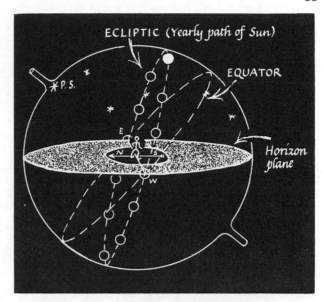

Fig. 14-2. Early Greek View

The Sun's yearly path through the star patterns was mapped. This is the tilted band called the ecliptic. The Sun is shown in one position (near mid-summer) and other positions are sketched. Here the celestial sphere is not spinning, but "frozen" with one star pattern overhead.

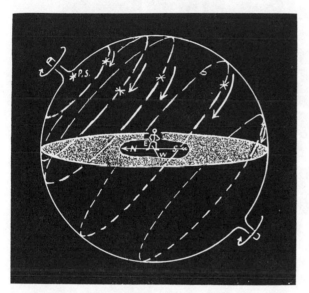

Fig. 14-1. The Universe According to Thales

the "first principle," the basic material of which all else is made. This was a bold beginning in "natural philosophy." He was a man of science, who *assumed that the whole universe could be explained by ordinary knowledge and reasoning.*

Others described the stars as set in a rotating sphere and discovered the obliquity of the ecliptic, the slant of the Sun's yearly path among the stars. This sorting out of the Sun's yearly motion from its daily motion was a useful step. The belt of star patterns along the Sun's yearly path came to be divided into twelve equal sections, the "signs of the zodiac," each named for a constellation. The Moon's path and the planets' paths are very near the Sun's, so they too travel through the signs of the zodiac.

Pythagoras (~ 530 b.c.). By the time Pythagoras established his school of philosophy—religion, sci-

ence, politics . . . —the time was ripe for the idea of a round Earth. Travellers' tales of ships and stars would suggest a curved Earth to an inquiring mind. Yet the picture of the Earth as a round ball is hard to believe. You accept it easily because you were indoctrinated when very young—but watch a child first learning about the antipodes where the people

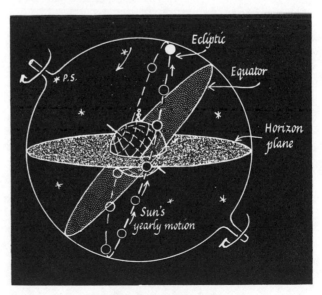

Fig. 14-3. Pythagorean View

The Pythagorean school adopted spherical Earth; and separated the general daily motion of stars, Sun, Moon, and planets, from the slow, backward motion of Sun, etc., through the star pattern.

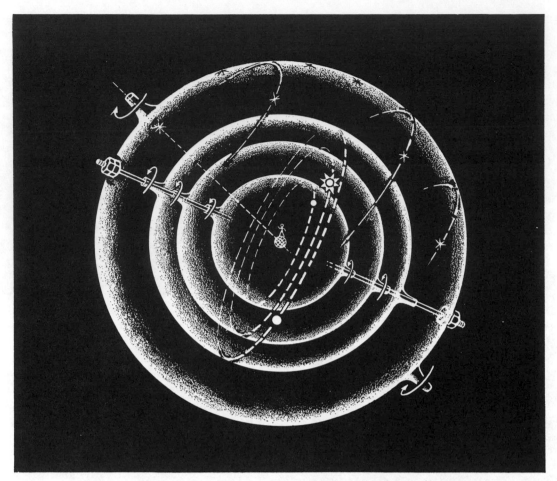

EARLY GREEK SYSTEM OF CRYSTAL SPHERES. (Pythagoras)

FIG. 14-4a. Part of the system, showing the rotating spheres of the Sun and two planets, carried around by the outer sphere of stars which spins daily.

are "upside down"! Pythagoras himself probably taught that the Earth is round; but we do not know whether most of the Pythagorean discoveries and views were his or those of later members of his school, which flourished for some 200 years. For the heavenly system, they pictured a round Earth inhabited all over, surrounded by concentric transparent spheres each carrying a heavenly body. The innermost sphere carried the Moon, obviously closer to the Earth than the rest. The outermost sphere carried the stars, and the intermediate ones carried Mercury, Venus, the Sun, Mars, Jupiter, and Saturn. The outer sphere of stars rotated once in a day and a night; the others ran slightly slower to show the lagging course of Sun, Moon, and planets. Here was a simple scientific theory, a conceptual scheme of rotating spheres that was simple (plain spheres, steady spins) and which could claim to be based on a simple general principle (spheres are the "perfect" shape and uniform rotation is the "perfect"

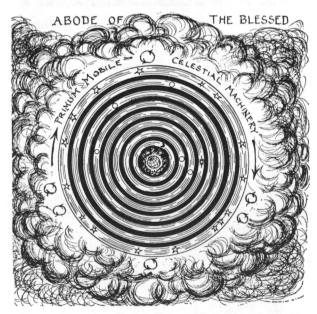

FIG. 14-4b. EARLY GREEK SYSTEM OF CRYSTAL SPHERES
A "section" of the whole system in the ecliptic plane.

motion for a sphere). The spheres carrying planets were arranged in order of their spinning speeds: Saturn, moving almost as fast as the stars—lagging only one revolution in thirty years—was placed just inside the stars; then Jupiter, Mars, Sun; with Venus and Mercury just inside or outside the Sun's sphere. This arranging by speeds was a lucky guess. We now know Saturn, Jupiter, and Mars are the "outer" planets, farther from the Sun than the Earth, with Saturn outermost, Mars nearest.

Some members of the school realized that a common 24-hour rotation could be separated, so they made the outer sphere of stars carry all the inner spheres with it. Then the inner ones had to spin slowly *backwards* within the outermost, thus carrying Sun, Moon, and planets backwards through the zodiac band of stars. Each inner sphere had its appropriate rate, once round in a year for the Sun, once in a month for the Moon, . . . once in 12 years for Jupiter. . . .

Pythagoras made discoveries in geometry. Though his "square on the hypotenuse" theorem was known long before him, he showed how to derive it. And he developed a theory of numbers. He preached that "numbers are the very essence of things," the basis of all natural knowledge; and his school was much concerned with the arithmetical properties of numbers and their use in science. He thus attached to the study of numbers mystical values that have appealed to thinking men from long before him till long after. Among primitive men, superstition gave lucky and unlucky numbers magic powers, and to this day reputable scientists discuss the structures of atoms and universes in terms of "magic numbers."[1] Such Pythagorean mysticism turns up again and again in the development of science. The hardheaded condemn it as a mischievous rock that can shipwreck rational science, but most of us welcome it as a lifebuoy that can keep fruitful speculation afloat when the way ahead seems stormy. The layman today finds it hard to distinguish between useful mysticism—such as dreams of a positive electron or of "anti-matter"— and cranky nonsense. But the difference is sharp enough: the modern scientist, even when he is being most mystical, uses a clear vocabulary of well-defined terms with agreed meaning between him and his colleagues; and he not only draws on experiment for suggestions and checks but insists on critical study of the reliability of the experimental evidence. The crank can quote experi-

ments to suit his purpose but fails to win confidence by his biased choice. There is, in fact, a corporate sanity among scientists that guides thinking in wise channels without restricting fruitful imagination.

Pythagoras was a sane scientist. In developing the science of music—a fine field in which to look for number properties—he assigned simple number ratios to musical harmonies. We keep these today: to be in perfect tune, two notes an octave apart must have vibration-frequencies that are as 2:1; two notes a musical fifth apart as 3:2. If different lengths of a harp string are chosen to give these harmonic intervals, they show the same proportions: the lengths are 2:1 for an octave, 3:2 for a fifth. Other simple ratios like 4:3 give a pleasant chord, but outlandish ones like 4.32 : 3.17 make an ugly dissonance in our ears trained by generations of the classical musical scale. This idea of ruling harmonic proportions was extended to astronomy. Pythagoras' pupils imagined the planetary spheres to be arranged by musical intervals: sizes and speeds had to fit simple number proportions. In rolling round with appropriate motion, each sphere made a musical note. The whole system of spheres made a harmony, "the music of the spheres," unheard by ordinary men, though some held that the master Pythagoras was privileged to hear it. Even this fanciful scheme was hardly unscientific, for its time. There was an utter absence of data; the distances of Sun and planets were unknown, and there was no prospect of measuring them; so the celestial harmonies merely added zest to thinking. A historian eight centuries later, wrote romantically: "Pythagoras maintained that the universe sings and is constructed in accordance with harmony; and he was the first to reduce the motions of the seven heavenly bodies to rhythm and song."[2]

Philolaus. The Sun, Moon, Venus, Mercury, Mars, Jupiter, Saturn—the seven planets as the Greeks listed them—all travel slowly *across the stars* from West to East. The star pattern carries the whole lot daily from East to West. This unfortunate reversal that spoiled the simplicity could be removed if the central Earth revolved instead of the stars; then all would revolve the same way. Philolaus, a pupil of Pythagoras, recorded such a view: instead of the Earth being the center of the Universe, there is a central fire—"the watchtower of the gods"—and the Earth swings round this fire daily in a small orbit, its inhabited part always facing outward away from the fire. This daily motion of the Earth would account for the daily motion of

[1] "Magic numbers" in nuclear physics are useful, so they are respectable. For an elementary modern example of almost meaningless number-witchcraft see the account of the Bohr atom in some beginning Chemistry texts.

[2] Hippolytos, quoted by G. Sarton in *A History of Science* (Harvard University Press, 1952), p. 214.

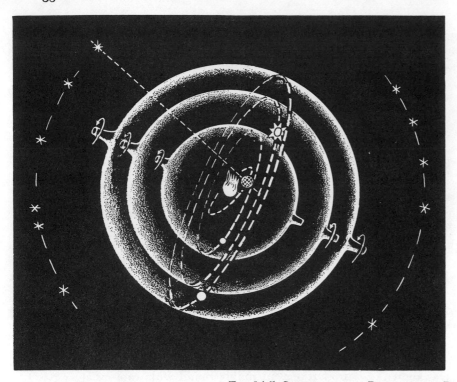
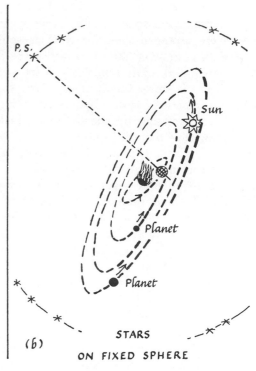

(b)

STARS
ON FIXED SPHERE

FIG. 14-5. SCHEME OF THE PYTHAGOREAN PHILOLAUS,
who pictured the Earth swinging around a central fire once in twenty-four hours.
This accounted for the *daily* motion of stars, Sun, Moon and planets.
Then spheres spinning slowly in the *same* direction carried the Moon, Sun, planets.
(a) View of spheres. (b) Skeleton scheme of orbits.

the stars across the sky: the outermost crystal sphere could be at rest. (Some went further and imagined an extra planet interposed between the Earth and the central fire. That counter-earth would protect the antipodes from being scorched—or perhaps it just *was* the antipodes—and it brought the total number of heavenly bodies up to the sacred Pythagorean number of ten.)

This fantastic scheme was revolutionary: it treated the Earth as a planet instead of making it the divine center, and it pointed out that the rotation of the starry sphere could be transferred to a daily revolution of the Earth. It may have paved the way for later theories of a moving Earth, but it did not last long, and it never suggested that the *Sun* is at the center nor even that the Earth is simply spinning. That last simplifying idea was suggested soon after, but it too did not find favor.

The Pythagoreans knew the Earth is round. They based their belief on a simple principle (spheres are perfect) and on practical facts. And they described the motions of the heavenly bodies by a rough but simple scheme that could be called a *theory*, in contrast with the more accurate workaday *rules* that were developed in Babylon. As a machine for making predictions, this first Greek system of

uniform spins was hopelessly inaccurate; but as a frame of knowledge it was indeed superior: it gave a feeling that the heavenly scheme of things makes sense.

Socrates (~ 430 B.C.). The great philosopher championed clear thinking with careful definitions, and condemned the astronomers for their wild conjectures. Thus he probably helped astronomy towards becoming an inductive science that extracts its picture from experimental observations.

At much the same time, two philosophers, *Democritus* and *Leucippus*, were constructing a theory of atoms to explain the properties of matter and even the structure of the world. It was unthinkable, they held, that matter could be chopped up into smaller and smaller pieces without limit. There must be tiny hard unsplittable atoms. Though they had no experimental evidence but only fanciful speculations, they managed to sketch a theory of fiery particles that looks sensible today.[3] They provided the idea

[3] It would be silly historical mysticism to call this Greek atomic theory, for all its modern flavor, a foreseeing of Dalton's atomic chemistry of 1800. It was not a scientific discovery 2000 years before its time. It was a great idea that had to wait 2000 years to direct the course of scientific thinking.

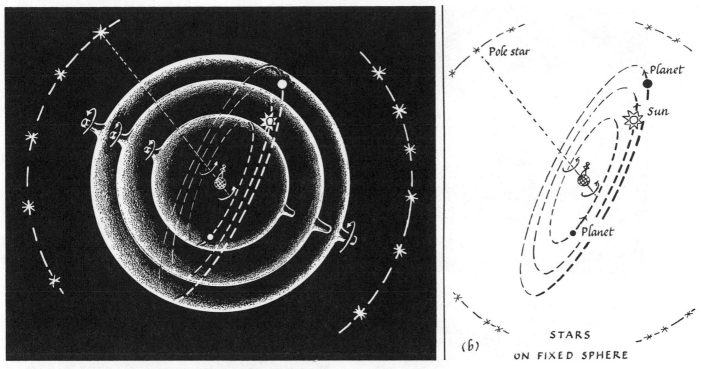

FIG. 14-6. LATER PYTHAGOREAN VIEW
The round Earth *spinning* accounted for the daily motion of stars, Sun, Moon and planets.
Then spheres rotating slowly in the same direction carried the Moon around once in a month,
the Sun around once in a year, and each planet around once in that planet's "year."
(a) View of spheres carrying Sun and two planets. (b) Skeleton scheme of orbits.

of atoms, to be mulled over, and occasionally used, through the centuries until chemical knowledge finally allowed an experimental atomic theory, in the last two hundred years. Their writings are lost, but the Latin poet Lucretius recorded their ideas in a magnificent poem two centuries later. He held that "reason liberates man from the terror of the gods"—a poetic version of the modern view "science cures superstition."

Though atomic theory did not concern astronomy directly, its insistence on a vacuum between atoms made it easier to think of empty space between heavenly bodies, or beyond them—contrary to the usual Greek view that space is limited and filled with an invisible æther.

Plato (~ 390 B.C.) was not much of an astronomer. He favored a simple scheme of spheres, and placed the planets in order of their speeds of revolution: Moon; Sun; Mercury and Venus travelling with the Sun; Mars; Jupiter; Saturn. The first great scheme that seemed to give a successful account of planetary motions came from Eudoxus, possibly at Plato's suggestion.

Eudoxus (~ 370 B.C.) studied geometry and philosophy under Plato, then travelled in Egypt, and returned to Greece to become a great mathematician and the founder of scientific astronomy. Gathering Greek and Egyptian knowledge of astronomy and adding better observations from contemporary Babylon, he devised a scheme that would save the phenomena.

The system of a few spheres, one for each moving body, was obviously inadequate. A planet does not move steadily along a circle among the stars. It moves faster and slower, and even stops and moves backwards at intervals. The Sun and Moon move with varying speeds along their yearly and monthly paths.[4] Eudoxus elaborated that scheme into a vast family of concentric spheres, like the shells of an onion. Each *planet* was given several adjacent spheres spinning about different axes, one within the next: 3 each for Sun and Moon, 4 each for the planets; and the usual outermost sphere of all for the stars. Each sphere was carried on an axle that ran in a hole in the next sphere outside it, and the axes of spin had different directions from one sphere to the next. The combined motions, with suitably chosen spins, imitated the observed facts. Here was a system that was simple in form (spheres) with a

[4] For example, the four seasons, from spring equinox to midsummer to autumn equinox, &c., are unequal. The Babylonians in their schemes for regulating the calendar by the new Moon had, essentially, time-graphs of the uneven motions of Moon and Sun.

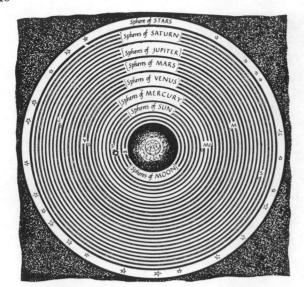

Fig. 14-7. Eudoxus' Scheme of Many Concentric Spheres
Each body, Sun, Moon or planet, had several spheres
spinning steadily around different axes. The
combination of these motions succeeded in imitating
the actual motions of Sun, Moon and even planets
across the star pattern.

simple principle (uniform spins), adjustable to fit the facts—by introducing more spheres if necessary. In fact, this was a good theory.

To make a good theory, we must have basic principles or assumptions that are simple; and we must be able to derive from them a scheme that fits the facts reasonably closely. Both the usefulness of a theory and our aesthetic delight in it depend on the simplicity of the principles as well as on the close fitting to facts. We also expect fruitfulness in making predictions, but that often comes with these two virtues of simplicity and accuracy. To the Greek mind, and to many a scientific mind today, a good theory is a simple one that can save all the phenomena with precision. Questions to ask, in judging a good theory, are, "Is it as simple as possible?" and "Does it save the phenomena as closely as possible?" If we also ask, "Is it *true*?", that is not quite the right requirement. We could give a remarkably *true* story of a planet's motion by just reciting its locations from day to day through the last 100 years; our account would be true, but so far from simple, so spineless, that we should call it just a list, not a theory.[5] The earlier Greek pictures with real crystal spheres had been like myths or tales for children— simple teaching from wise men for simple people. But Eudoxus tried to devise a successful machine that would express the actual motions and predict

their future. He probably considered his spheres geometrical constructions, not real globes, so he had no difficulty in imagining several dozen of them spinning smoothly within each other. He gave **no** mechanism for maintaining the spins—one might picture them as driven by gods or merely imagined by mathematicians.

Here is how Eudoxus accounted for the motion of a planet, with four spheres. The planet itself is carried by the innermost, embedded at some place on the equator. The outermost of the four spins round a North-South axle once in 24 hours, to account for the planet's daily motion in common with the stars. The next inner sphere spins with its axle pivoted in the outermost sphere and tilted 23½° from the N-S direction, so that *its* equator is the ecliptic path of the Sun and planets. This sphere revolves in the planet's own "year" (the time the planet takes to travel round the zodiac), so its motion accounts for the planet's general motion through the star pattern.[6] These two spheres are equivalent to two spheres of the simple system, the outermost sphere of stars that carried all the inner ones with it, and the planet's own sphere. The third and fourth spheres have equal and opposite spins about axes inclined at a small angle to each other. The third sphere has its axle pivoted in the zodiac of the second, and the fourth carries the planet itself em-

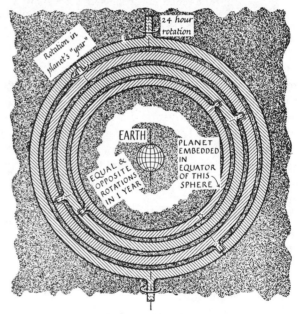

Fig. 14-8. Part of Eudoxus' Scheme:
Four Spheres to Imitate the Motion of a Planet
The sketch shows machinery for one planet.
The outermost sphere spins once in twenty-four hours.
The next inner sphere rotates once in the planet's
"year." The two innermost spheres spin with equal
and opposite motions, once in our year, to produce
the planet's epicycloid loops.

---

[5] Young scientists are urged, nowadays, not to be satisfied with just collecting specimens, or facts or formulas, lest they get stuck at the pre-Greek stage.

[6] In terms of our view today, the spin of the outermost sphere corresponds to the Earth's daily rotation; the spin of the next sphere corresponds to the planet's own motion along its orbit round the Sun; the spins of the other two spheres combine to show the effect of viewing from the Earth which moves yearly around the Sun.

EVIDENCES FOR ROUND EARTH

ANCIENT

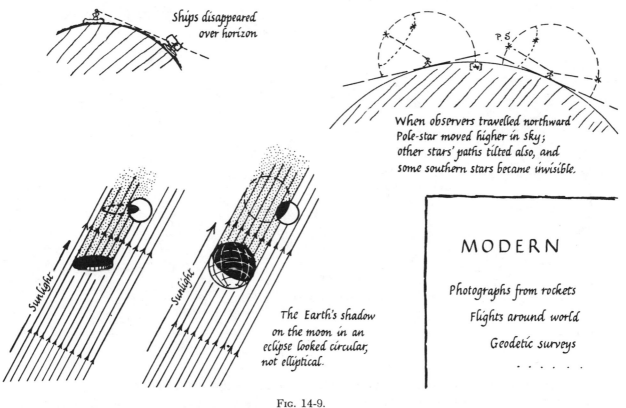

Ships disappeared over horizon

When observers travelled northward Pole-star moved higher in sky; other stars' paths tilted also, and some southern stars became invisible.

The Earth's shadow on the moon in an eclipse looked circular, not elliptical.

MODERN

Photographs from rockets

Flights around world

Geodetic surveys

. . . . . .

FIG. 14-9.

bedded in the equator. Their motions combine to add the irregular motion of stopping and backing to make the planet follow a looped path. The complete picture of this three dimensional motion is difficult to visualize.

With 27 spheres in all, Eudoxus had a system that imitated the observed motions quite well: he could save the phenomena. The basis of his scheme was simple: perfect spheres, all with the same center at the Earth, spinning with unchanging speeds. The mathematical work was far from simple: a masterpiece of geometry to work out the effect of four spinning motions for each planet and choose the speeds and axes so that the resultant motion fitted the facts. In a sense, Eudoxus used harmonic analysis—in a three-dimensional form!—two thousand years before Fourier. It was a good theory.

But not very good: Eudoxus knew there were discrepancies, and more accurate observations revealed further troubles. The obvious cure—add more spheres—was applied by his successors. One of his pupils, after consulting Aristotle, added 7 more spheres, greatly improving the agreement. For example, the changes in the Sun's motion that make the four seasons unequal were now predicted

properly. Aristotle himself was worried because the complex motion made by one planet's quartet of spheres would be handed on, unwanted, to the next planet's quartet. So he inserted extra spheres to "unroll" the motion between one planet and the next, making 55 spheres in all. The system seems to have stayed in use for a century or more till a simpler geometrical scheme was devised. (An Italian enthusiast attempted to revive it 2000 years later, with 77 spheres.)

Aristotle (340 B.C.), the great teacher, philosopher, and encyclopedic scientist was the "last great speculative philosopher in ancient astronomy." He had a strong sense of religion and placed much of his belief in the existence of God on the glorious sight of the starry heavens. He delighted in astronomy and gave much thought to it. In supporting the scheme of concentric spinning spheres, he gave a dogmatic reason: *the sphere is the perfect solid shape*; and this prejudiced astronomical thinking about orbits for centuries. By the same token, the Sun, Moon, planets, stars must be spherical in form. The heavens, then, are the region of perfection, of unchangeable order and circular motions. The space

EVIDENCES FOR SPINNING EARTH

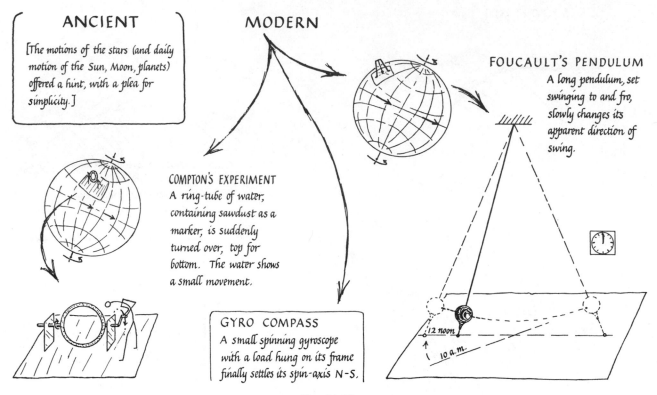

ANCIENT

[The motions of the stars (and daily motion of the Sun, Moon, planets) offered a hint, with a plea for simplicity.]

MODERN

COMPTON'S EXPERIMENT
A ring-tube of water, containing sawdust as a marker, is suddenly turned over, top for bottom. The water shows a small movement.

GYRO COMPASS
A small spinning gyroscope with a load hung on its frame finally settles its spin-axis N–S.

FOUCAULT'S PENDULUM
A long pendulum, set swinging to and fro, slowly changes its apparent direction of swing.

12 noon

10 a.m.

FIG. 14-10.

between Earth and Moon is unsettled and changeable, with vertical fall the natural motion.

For ages Aristotle's writings were the only attempt to systematize the whole of nature. They were translated from language to language, carried from Greece to Rome to Arabia, and back to Europe centuries later, to be copied and printed and studied and quoted as the authority. Long after the crystal spheres were discredited and replaced by eccentrics, those circles were spoken of as spheres; and the medieval schoolmen returned to crystal spheres in their short-sighted arguments, and believed them real. The distinction between the perfect heavens and the corruptible Earth remained so strong that Galileo, 2000 years later, caused great annoyance by showing mountains on the Moon and claiming the Moon was earthy; and even he, with his understanding of motion, found it hard to extend the mechanics of downward earthly fall to the circular motion of the heavenly bodies.

Aristotle made a strong case for the Earth itself being round. He gave theoretical reasons:

   (i) *Symmetry*: a sphere is symmetrical, perfect

   (ii) *Pressure*: the Earth's component pieces, falling *naturally* towards the center, would press into a round form

and experimental reasons:

   (iii) *Shadow*: in an eclipse of the Moon, the Earth's shadow is always circular: a flat disc could cast an oval shadow.

   (iv) *Star heights*: even in short travels Northward or Southward, one sees a change in the position of the star pattern.

This mixture of dogmatic "reasons" and experimental common sense is typical of him, and he did much to set science on its feet. His whole teaching was a remarkable life work of vast range and enormous influence. At one extreme he catalogued scientific information and listed stimulating questions; at the other extreme he emphasized the basic problems of scientific philosophy, distinguishing between the *true physical causes* of things and *imaginary schemes to save the phenomena*.

Euclid, soon after Aristotle, collected earlier discoveries in geometry, added his own, and produced his magnificent science: geometry developed by deductive logic. Such mathematics is automatically true to its own assumptions and definitions. Whether it also fits the natural world is a matter for experiment. Therefore, we should neither question the truth of a piece of mathematics, nor call it a *natural* science.

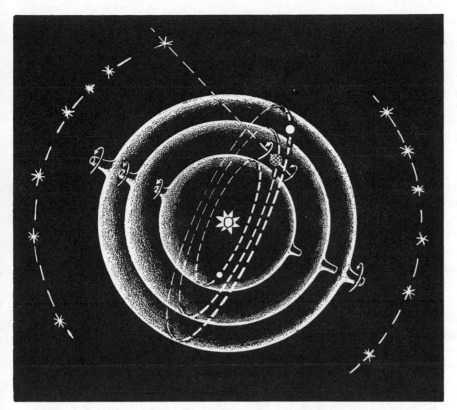
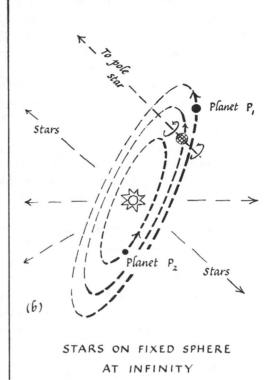

STARS ON FIXED SPHERE
AT INFINITY

FIG. 14-11. ARISTARCHUS' SCHEME
Only two specimen planets are shown. P₁ might be Mars, Jupiter or Saturn.
P₂ might be Mercury or Venus.
(a) View of spheres. (b) Skeleton scheme showing planetary orbits.

The Scientific School at Alexandria

Alexander the Great built a huge empire, sweeping in a dozen years from Greece through Asia Minor, Egypt, Persia, to the borders of India and back to Babylon. Early in his campaign, he founded the great city of Alexandria at the mouth of the Nile. Greek scholars collected there, and the Museum or University of Alexandria grew to be a great center of learning. A school of astronomers started there about 330 B.C. and flourished for centuries. Some made new and more accurate observations, devising new instruments; some made a new kind of advance by trying to measure the actual sizes and distances of the Sun and Moon; and some produced new and better theories.

Before the new school changed the spinning spheres into eccentric circles, one Greek astronomer, *Aristarchus* (∼ 240 B.C.), made two simplifying suggestions:

(i) *The Earth spins*—and *that* accounts for the daily motion of the stars. (Others had made this suggestion.)

(ii) *The Earth moves round the Sun in a yearly orbit*; and the other planets do likewise—that accounts for the apparent motions of the Sun and planets across the star patterns.

This simple scheme failed to catch on: tradition was against it; and it was merely an idea, not backed by a reasoned set of measurements, such as Copernicus gave many centuries later. Earth moving around an orbit raised mechanical objections that seemed even more serious in later ages; and it raised a great astronomical difficulty immediately. If the Earth moves in a vast orbit, the pattern of fixed stars should show parallax changes during the year. None were observed, and Aristarchus could only reply that the stars must be almost infinitely far off compared with the diameter of the Earth's orbit. Thus he pushed the stars away to far greater distances. He also released them from being all on one great sphere. As long as they are far enough away, they may be scattered through a great range of space, at rest while the Earth spins.

44

Measurements of Size and Distance

Astronomers were now trying to gauge the actual sizes of Sun, Moon, and Earth, and their distances apart. Earlier, there had been vague guesses: the Sun and Moon are very far away, or they are only just beyond the clouds; the Sun is the size of Greece, the Moon smaller. . . . Definite measurements would turn astronomy into a much more real science, but they were difficult to make.

Man judges ordinary distances by his eye-muscles, estimating the angle of squint when both eyes are directed on the object. For remote objects, our eyes are too close, and we use a longer base line and actually measure angles. Then we draw a diagram to scale, or use trigonometry. We now know that, for the Moon, a base line of 1000 miles gives an angle-of-squint of only ¼°. And for the Sun it would be only 1/1600°, a very difficult difference to measure even today, with observers so far apart.

The *size* of the Sun (or Moon) is connected with its *distance* from us by an easy measurement: its *angular diameter*. Fix a coin on a window and move until it just blots out the Sun's disc. Measure the coin's diameter and distance from your eye: their ratio, coin diameter/coin distance, gives the ratio, Sun's diameter/Sun's distance. The ratio is about 1/110. Or use some instrument to measure the angle the Sun's diameter subtends—almost exactly ½°. Draw a triangle with vertex angle ½° on a big sheet of paper and measure its sides. Or use simple trigonometry. You will find the distance from base to vertex is about 110 times the base. *The Sun's distance from us is 110 times its diameter.* Almost the same proportion holds for the Moon—Moon and Sun *look* about the same size, and this is confirmed in total eclipses of the Sun, when the Moon only just

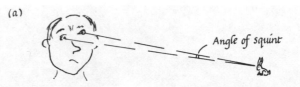

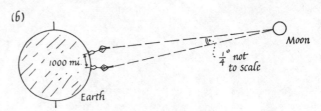

FIG. 14-13. ESTIMATING DISTANCES
(a) Judging distance by squint.
(b) Judging Moon's distance by squint would require observers 1000 miles apart to notice a difference of ¼°.

manages to cover it. A measurement of *one* of the two quantities DIAMETER and DISTANCE then combines with 110 to give the other. The usual measurement is DISTANCE, estimated by squint.

Size of the Earth

The first measurement to be made was the size of the Earth itself, and the other measurements emerged in terms of the Earth's radius.

Eratosthenes (~ 235 B.C.) made one of the early estimates. He compared the direction of the local vertical with parallel beams of sunlight at two stations a measured distance apart. He assumed that the Sun is so remote that all sunbeams reaching the Earth at any instant are practically parallel.

He needed *simultaneous* observations at two stations far apart. Good clocks that could be compared and transported were not available. So he obtained

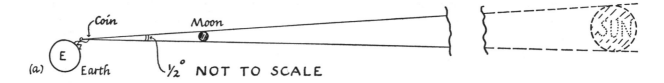

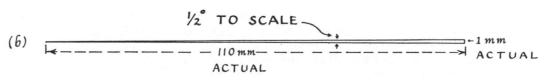

FIG. 14-12. RELATION BETWEEN SIZE AND DISTANCE can be found by holding measured coin at measured distance. This does not tell us absolute size or distance.
Sketch (a) is *not* to scale. Sketch (b) shows the "angular size" of Sun and Moon *drawn to scale*.
Measurements show the Sun and Moon each subtend about ½° at Earth. Measurements or trigonometry tables give a proportion of about 1:110 for base:height.

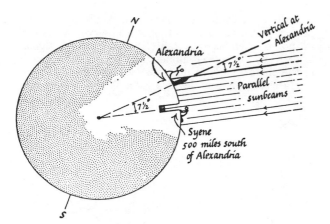

FIG. 14-14. How Eratosthenes Estimated
the Size of the Earth

simultaneity by choosing noon (highest Sun) on the same day at stations in the same longitude. He used observations at Alexandria, where he worked, and at Syene,[7] 500 miles farther south. The essential observation at Syene was this: at noon on midsummer day, June 22, sunbeams falling on a deep well there reach the water and are reflected up again. Eratosthenes knew this from library information. Therefore the noonday Sun must be vertically overhead at Syene on that day. At noon on the same day of the year, he measured the shadow of a tall obelisk at Alexandria and found that the Sun's rays made 7½° with the vertical. He assumed that all sunbeams reaching the Earth are parallel. So it was the vertical

[7] Modern name: Aswân, where the great dam has been built on the Nile.

(the Earth's radius) that had different directions. Therefore the Earth's radii to Alexandria and Syene make 7½° at the center. Then if 7½° carries 500 miles of Earth's circumference, 360° carries how much? The rest was arithmetic. Measuring the 500 miles separation was hard—probably a military measurement done by professional pacers. There is doubt about the units he used, but some say his error was less than 5%—a remarkable success for this early simple attempt. He also guessed at the distances of Sun and Moon.

Moon's Size and Distance

The size of the Moon was compared with the size of the Earth by watching eclipses of the Moon. Timing the Moon as it crossed the Earth's shadow, Aristarchus found that the diameter of the round patch of Earth's shadow out at the Moon was 2½ Moon diameters. If the Sun were a point-source of light at infinity, it would shadow the Earth in a parallel beam as wide as the Earth itself. In that case we should have:

EARTH'S DIAMETER = 2½ MOON DIAMETERS;
or MOON'S DIAMETER = ⅖ EARTH'S DIAMETER,
∴ MOON'S DISTANCE, which is 110 MOON DIAMETERS
    = (⅖) 110 EARTH DIAMETERS
    = 44 EARTH DIAMETERS or 88 EARTH RADII.

That, with Eratosthenes' value of about 4000 miles for the Earth's radius, would place the Moon 350,000 miles away. Taking the Sun at infinity is reasonable. Treating it as a point source is not, and of course

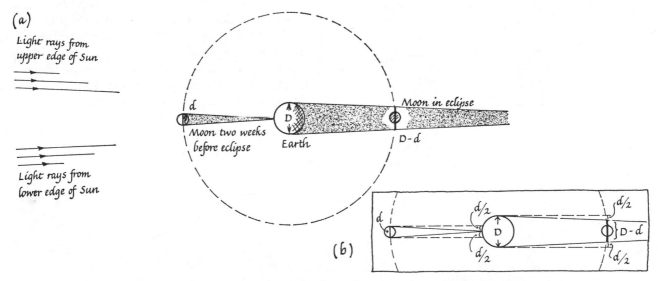

FIG. 14-15. Early Greek Measurement of Size of the Moon (and Therefore Its Distance).
Observations of eclipses showed that the width of the Earth's shadow at the Moon is 2.5 Moon-diameters. However, the Earth's shadow narrows as its distance from Earth increases because the Sun is not a point-source. Since the Moon's shadow almost dies out in the Moon-Earth distance, the Earth's shadow must narrow by the same amount—one Moon-diameter—in the same distance. Then Earth-diameter must be 3.5 Moon-diameters.

Aristarchus knew that. The Sun is a great flaming globe that casts tapering shadows (of angle about ½°). When there is a total eclipse of the *Sun* the Moon can only just shut the whole Sun off from our eyes; so the Moon's shadow cone only just reaches us—it ends practically at the Earth. Therefore, in running the Moon-to-Earth distance, the Moon's shadow narrows by the whole Moon diameter. In an eclipse of the *Moon*, the *Earth's* shadow is thrown the same distance, Earth-to-Moon, so it too must lose the same amount of width, one Moon diameter. Thus, Aristarchus argued:

EARTH'S DIAMETER — one MOON DIAMETER
$\qquad$ = 2½ MOON DIAMETERS,

∴ EARTH'S DIAMETER = (1 + 2½) MOON DIAMETERS
$\qquad$ = 3½ MOON DIAMETERS,

∴ MOON'S DISTANCE = 110 MOON DIAMETERS
$\qquad$ = 2/7 (110) EARTH DIAMETERS
$\qquad$ = 31.4 EARTH DIAMETERS or 63 EARTH RADII.

More accurate measurements, used by Aristarchus and his successors, gave 60 Earth's radii (within 1% of the modern measurement), about 240,000 miles.

Later, the Moon's distance was measured by the squint method, thus: Observers at two stations far apart, in the same longitude, sight the Moon at the same time on an agreed night. They measure the angle between the *Moon's direction* and a *vertical plumb-line* (which gives the local zenith direction). These angles u, v suffice to locate the Moon if the base-line distance between the stations is known. With stations far apart, that distance would be a difficult measurement for early astronomers; but the angle between the Earth's radii to the two stations will do instead. So the observer at each station measures the angle between his local vertical

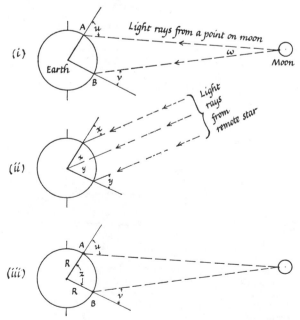

FIG. 14-16. MEASURING MOON'S DISTANCE BY SQUINT METHOD

and the light from a standard star. Any star will do: a perfect pole star, or any other star observed at its highest point. Then, as in sketch (ii) of Fig. 14-16, the sum of these two measured angles (x + y) gives the angle z at the center. Now in diagram (iii) the three angles u, v, z are known, and the two radii R and R are known

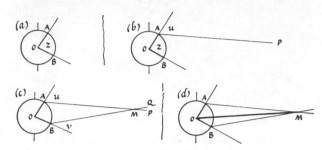

FIG. 14-17. SCALE DIAGRAM Method for Calculating (MOON'S DISTANCE)/(EARTH'S RADIUS) from measurements

to be equal. To find the Moon's distance, either use trigonometry or make a simple drawing to scale, thus: on a big sheet of paper (sand on the floor for early astronomers), draw a circle and mark radii OA and OB making the angle z (known by adding measurements x + y). Continue these radii out to represent the zenith verticals at stations A and B. From A draw a "Moon-line" AP making measured angle u with the radius there. Draw BQ from B. Where they cross is the Moon's position M on the scale diagram. Measure OM and divide it by the radius OA. The result gives the Moon's distance as a multiple of Earth's radius.

Accurate measurements give

MOON'S DISTANCE = just over 60 EARTH'S RADII
$\qquad$ ≈ 240,000 miles

Sun's Size and Distance

The Sun's distance is much harder to estimate even today because the Sun is in fact so far away, so bright, and so big. The squint method angle at the Sun is too small to measure without telescopic accuracy. However Aristarchus used an ingenious scheme to get a rough estimate. He watched the Moon for the stage at which it showed *exactly* half moon. Then sunlight must be falling on the Moon at right angles to the observer's line of sight, EM. At that instant, he measured the angle between the

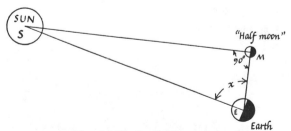

FIG. 14-18. SUN'S DISTANCE
Early Greek estimate of the Sun's distance from the Earth, in terms of the Moon's known distance. Greek astronomers tried to measure the angle x (or SEM), which is itself nearly 90°.

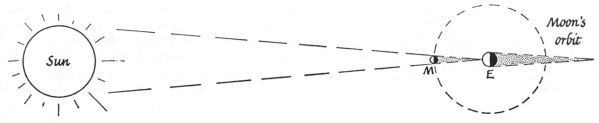

Fig. 14-19a. Sun, Moon, Earth. Sketch *not* to scale. The Sun is shown much too near, and the Moon is too big and too near.

Moon's direction and the Sun's direction. This angle, SEM, was nearly a right angle but not quite. Then in the great triangle, SEM, two angles were known. The third angle, ESM, is the small one that gives the essential measure of the Sun's distance. It is got by subtracting from 180°. It is small: 3° as Aristarchus estimated, but really only ⅙°. So Aristarchus' conclusion that the Sun's distance is about 20 times the Moon's was an underestimate, about 20 times too small. This proportion, sun's distance/moon's distance, is got from the angles by a scale drawing or by very simple trigonometry. (EM/ES is the cosine of the measured angle SEM. Therefore $\dfrac{ES}{EM} = \dfrac{1}{\cos SEM}$, easily taken from trig. tables.)

Thus, the astronomers at Alexandria had estimates of the dimensions of the heavenly system, and these were used by their successors with little change for many centuries:

EARTH: radius 4000 miles

MOON: distance from Earth, 60 Earth's radii or 240,000 miles
personal radius 1100 miles

SUN: distance from Earth ?? 1200 Earth's radii. (Thought to be inaccurate: it was.)
personal radius ?? 44,000 miles

PLANETS: distances quite unknown, but presumed farther than Moon

STARS: distances quite unknown, presumed beyond Sun and planets

Looking at these estimates, you can see that the usual pictures in books to illustrate eclipses are badly out of scale. Fig. 14-19 shows more realistic pictures based on modern measurements. No wonder eclipses are rare. It is easy enough to miss the slender cones of shadow. The Moon's orbit is tilted about 5° from the Sun's apparent path, and that makes eclipses still rarer.

New Theories: Eccentrics; Epicycles

In the school at Alexandria, the bold suggestion of making the Earth spin and move round a central Sun did not find favor. A stationary central Earth remained the popular basis, but spinning concentric

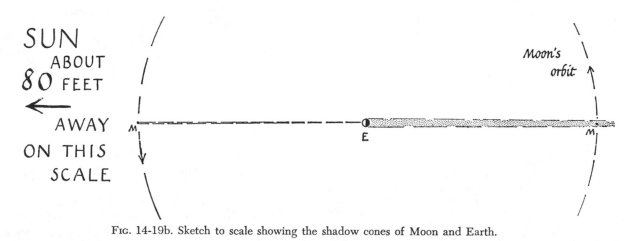

Fig. 14-19b. Sketch to scale showing the shadow cones of Moon and Earth.

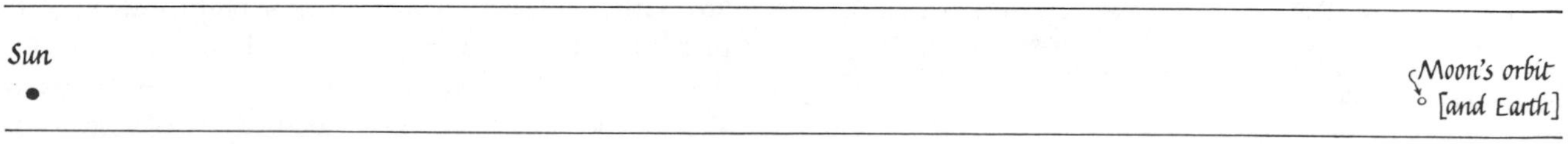

Fig. 14-19c. Sketch to scale. Here the scale has been reduced so that Sun, Moon and Earth are in the picture.
The small circle is the Moon's orbit. The Earth, at the center of that circle, is too small to show.
On this scale it is a dot ¹⁄₁₀₀₀ inch in diameter. The Moon is much too small to show.

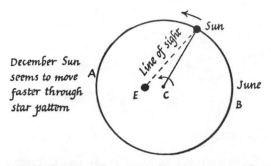

FIG. 14-20. THE ECCENTRIC SCHEME FOR THE SUN
The Sun is carried around a circular path by a radius
that rotates at constant speed, as in the simplest
system of spheres. The observer, on the Earth, is
off-center, so that he sees the Sun move unevenly—
as it does—faster in December, slower in June.

spheres made the model too difficult. Instead, the
slightly uneven motion of the Sun around its "orbit"
could be accounted for by a single eccentric circle:
the Sun was made to move steadily round a circle,
with the Earth fixed a short distance off center.
Then, as seen from the Earth, the Sun would seem
to move faster at some seasons (about December,
at **A**) and slower 6 months later (at **B**). This was
still good theory. For good theory, the scheme
should have an appealing simplicity and be based
on simple assumptions.[8] These needs were met: a
perfect circle of constant radius, and motion with
constant speed round it. Such constancies were necessary to the Greek mind—in fact to any orderly
scientific mind. Without them theory would de-
generate into a pack of demons. Placing the Earth

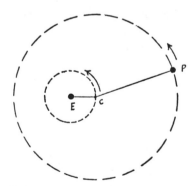

FIG. 14-21. THE ECCENTRIC SCHEME, FOR A PLANET
Each planet is carried at the end of a radius that rotates
at constant speed; but this whole circle—center, radius
and planet—revolves once a year round the eccentric
Earth. (To picture the motion, imagine the radius CP
continued out to be the handle of a frying pan. Then
imagine the frying pan given a circular motion (of
small radius, EC) around E as center, by a housewife
who wants to melt a piece of butter in it quickly. Then
make the handle CP revolve too—very slowly, for an
outer planet like Jupiter.)

---

[8] The assumptions should be *logical* in the schoolboy slang
sense—a sense that would have pained the Greek thinkers.
In *their* use, it was the argument from the assumptions that
had to be logical.

off-center was a regrettable lapse from symmetry,
but then the Sun's speed *does* behave unsymmetri-
cally—our summer is longer than our winter. A
similar scheme served for the Moon, but the planets
needed more machinery. Each planet was made to
move steadily around a circle, once around in its
own "year," with the Earth fixed off center; but then
the whole circle, *planet's orbit and center* was made
to revolve around the Earth once in 365 days. This
added a small circular motion (radius EC) to the
large main one, producing the planet's epicycloid
track. The daily motion with the whole star pattern
was superimposed on this.

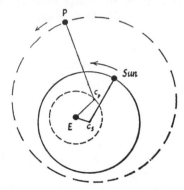

FIG. 14-22. THE ECCENTRIC SCHEME
Sketch showing machinery for motions of Sun and one planet.

Another scheme to produce the same effect used
a fixed main circle (the deferent) with a radius arm
rotating at constant rate. The end of that arm car-

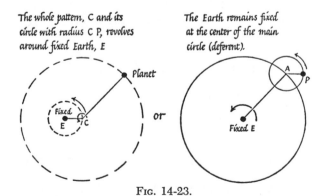

FIG. 14-23.
THE ECCENTRIC SCHEME AND THE EPICYCLE SCHEME

ried a small sub-circle (the epicycle). A radius of
that small circle carried the planet round at a steady
rate, once in 365 days. Though these schemes oper-
ate with circles, they were described more grandly
in terms of spinning spheres and sub-spheres. For
many centuries, astronomers thought in terms of
such "motions of the heavenly spheres"—the
spheres themselves growing more and more real as
Greek delight in pure theory gave place to childish
insistence on authoritarian truth.

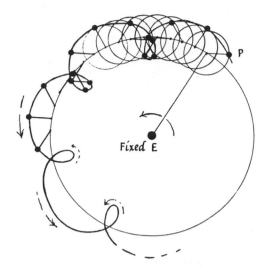

FIG. 14-24.
MAKING THE PATH OF A PLANET BY THE EPICYCLE SCHEME.
This sketch shows how the two circular motions combine
to produce the epicycloid pattern that is observed
for a planet.

Hipparchus (∼ 140 B.C.), "one of the greatest mathematicians and astronomers of all time,"[9] made great advances. He was a careful observer, made new instruments and used them to measure star positions. A new star that blazed into prominence for a short time is said to have inspired him to make his great star catalogue, classifying stars by brightness and recording the positions of nearly a thousand by celestial latitude and longitude. He made the first recorded celestial globe. There were no telescopes,[10] nothing but human eyes looking through sighting holes or along wooden sticks. Simple instruments like dividers measured angles between one star (or planet) and another, or between star and plumb line. Yet Hipparchus tried to measure within ⅙°.

He practically invented spherical trigonometry for use in his studies of the Sun and Moon. He showed that eccentrics and epicycles are equivalent for representing heavenly motions. Adding his own observations to earlier Greek ones and Babylonian records, he worked out epicyclic systems for the Sun and Moon. The planets proved difficult for want of accurate data, so he embarked on new measurements.

From Greek observations dating back 150 years Hipparchus discovered a very small but very important astronomical creep: the "precession of the equinoxes." At the spring equinox, midway between winter and summer, the Sun is at a definite place

in the zodiac, and it returns there each year. But Hipparchus discovered that the Sun is not quite in the same patch of stars at the next equinox. In fact it gets around to the old patch of stars about 20 minutes too late, so that at the exact instant of equinox it is a little earlier on its zodiac path, about ¹⁄₇₀° after one year, nearly 1½° after a century. Hipparchus found this from changes in star longitudes between old records and new. Longitudes are reckoned along the zodiac, from the spring equinox where the equator cuts the ecliptic. Since all longitudes seemed to be changing by one or two degrees a century, Hipparchus saw that the zodiac girdle must be slipping round the celestial sphere at this rate, carrying all the stars with it, leaving the celestial equator fixed with the fixed Earth.[11] This motion seems small—a whole cycle takes 26,000 years— yet it matters in astronomical measurements, and has always been allowed for since Hipparchus discovered it. The discovery itself was a masterpiece of careful observing and clever thinking.

Precession remained difficult to visualize until Copernicus, sixteen centuries later, simplified the story with a complete change of view (see Ch. 16). Even then it remained unexplained—unconnected with other celestial phenomena—until Newton gave a simple explanation. Discovered as a mysterious creep, precession is now a magnificent witness to gravitation.

Hipparchus left a fine catalogue of stars, epicycle schemes, and good planetary observations—a magnificent memorial to a great astronomer. These had to wait two and a half centuries for the great mathematician Ptolemy to organize them into a successful theory.

---

[9] Sarton, *History of Science.*

[10] That invention was seventeen centuries in the future. By magnifying a patch of sky it enables much finer measurements to be made.

[11] The Sun's ecliptic path cuts the celestial equator in two points. When the Sun reaches either of these it is symmetrical with respect to the Earth's axis. Day and night are equal for all parts of the Earth: that is an equinox. Precession is a slow rotation of the whole celestial sphere, including the zodiac and the Sun, around the *axis of the ecliptic*, perpendicular to the ecliptic plane. From one century to the next, the creeping of precession brings a slightly different part of the zodiac belt to the equinox-points (*where ecliptic cuts equator*)—hence the name. The whole celestial sphere joins in this slow rotation round the ecliptic axis. This applies, for example, to the stars near the N-S axis, which is fixed with the Earth and 23½° from the ecliptic axis; so the motion carries the current pole-star away from the N-S axis and brings a new one in the course of time. Thus, in some centuries there is a bright star in the right position for pole-star, and in others there is no real star, only a blank in the pattern. In the 40-odd centuries between the building of the pyramids and the present, this motion has accumulated a considerable effect. In fact, by examining the pyramid tunnels that were built to face the dog-star Sirius at midnight at the Spring equinox, we can guess roughly how long ago they were built.

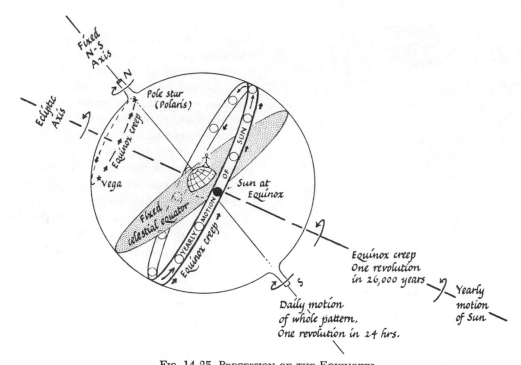

Fixed N-S Axis

N

Ecliptic Axis

Pole star (Polaris)

Equinox creep

Vega

Sun at Equinox

Fixed celestial equator

YEARLY MOTION OF SUN

Equinox creep

S

Daily motion of whole pattern, One revolution in 24 hrs.

Equinox creep One revolution in 26,000 years

Yearly motion of Sun

FIG. 14-25. PRECESSION OF THE EQUINOXES

In addition to (a) the daily motion of the whole heavens around the N-S axis fixed in the fixed Earth and (b) the yearly motion of the Sun around its ecliptic path in the zodiac band of stars, Hipparchus discovered (c) a slow rotation of the whole pattern of stars around a different axis, *the ecliptic axis* (perpendicular to the zodiac belt).

Ptolemy (~A.D. 120) made a "critical reappraisal of the planetary records." He collected the work of Hipparchus and his predecessors, adding his own observations, evolved a first-class theory and left a masterly exposition that dominated astronomy for the next fourteen centuries. The positions of the Sun, Moon, and planets, relative to the fixed stars, had been mapped with angles measured to a fraction of a degree. He could therefore elaborate the system of eccentric crystal spheres and epicycles and refine its machinery, so that it carried out past motions accurately and could grind out future predictions with success. He devised a brilliant *mathematical machine*, with simple rules but complex details, that could "save the phenomena" with century-long accuracy. In this, he neglected the crystal spheres as moving agents; he concentrated on the rotating spoke or radius that carried the planet around, and he provided sub-spokes and arranged eccentric distances. He expounded his whole system for Sun, Moon, and planets in a great book, the *Almagest*.

Ptolemy set forth this general picture: the heaven of the stars is a sphere turning steadily round a fixed axis in 24 hours; the Earth must remain at the center of the heavens—otherwise the star pattern would show parallax changes; the Earth is a sphere, and it must be at rest, for various reasons—e.g., objects

thrown into the air would be left behind a moving Earth. The Sun moves round the Earth with the simple epicycle arrangement of Hipparchus, and the Moon has a more complicated epicyclic scheme.

In his study of the "five wandering stars," as he called the planets, Ptolemy found he could not "save the phenomena" with a simple epicyclic scheme. There were residual inequalities or discrepancies between theory and observation. He tried an epicycle scheme with the Earth eccentric, moved out from the center of the main circle. That was not sufficient, so he not only moved the Earth off center but also moved the center of uniform rotation out on the other side. He evolved the successful scheme shown in Fig. 14-26. C is the center of the main circle; E is the eccentric Earth; Q is a point called the "equant," an equal distance the other side of C (QC = CE). An arm QA rotates with constant speed around Q, swinging through equal angles in equal times, carrying the center of the small epicycle, A, round the main circle. Then a radius, AP, of the epicycle rotates steadily, carrying the planet P. It was a desperate and successful attempt to maintain a scheme of circles, with constant rotations. The arm of the little circle carrying the planet rotated at constant rate. Ptolemy felt compelled to have an arm of the main circle also rotating at constant rate. To fit the facts, that arm could not be a

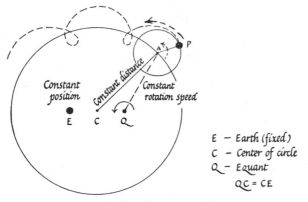

FIG. 14-26. THE PTOLEMAIC SCHEME
This system imitated the motions of
Sun, Moon, and planets very accurately.

E — Earth (fixed)
C — Center of circle
Q — Equant
QC = CE

In the *Almagest*, Ptolemy described a detailed scheme for each planet and gave tables from which the motion of each heavenly body could be read off. The book was copied (by hand, of course), translated from Greek to Latin to Arabic and back to Latin as culture moved eastward and then back to Europe. There are modern printed versions with translations. It served for centuries as a guide to astronomers, and a handbook for navigators. It also provided basic information for that extraordinary elaboration of man's fears, hopes, and greed—astrology—which needed detailed records of planetary positions.

The Ptolemaic scheme was efficient and intellectually satisfying. We can say the same of our modern atomic and nuclear theory. If you asked whether it is true, both Greeks and moderns would question your word "true"; but if you offered an alternative that is simpler and more fruitful, they would welcome it.

radius from the center of the main circle, as in the simplest epicycle scheme. Nor could it be the arm from an eccentric Earth E. But he could save the phenomena with an arm from the equant point Q that did rotate at constant speed. Thus for each planet's main circle, there were three points, all quite close together, each with a characteristic constancy.

E the Earth in fixed position	C the center of the main circle with arm CA of constant length	Q the equant point with arm QA rotating at constant speed

By choosing suitable radii, speeds of rotation, and eccentric distance EC (= CQ), Ptolemy could save the phenomena for all planets (though Mercury required a small additional circle). The main circle was given a different tilt for each planet, and the epicycle itself was tilted from the main circle.

Here was a gorgeously complicated system of main circles and sub-circles, with different radii, speeds, tilts, and different amounts and directions of eccentricity. The system worked: like a set of mechanical gears, it ground out accurate predictions of planetary positions for year after year into the future, or back into the past. And, like a good set of gears, it was based on essentially simple principles: circles with *constant radii*, rotations with *constant speeds, symmetry* of equant (QC = CE), *constant* tilts of circles, and the Earth fixed in a constant position.[12]

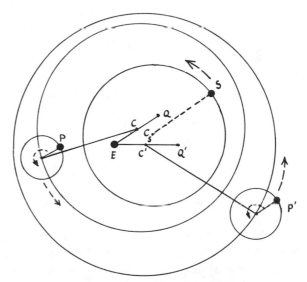

FIG. 14-27. THE PTOLEMAIC SCHEME
for the Sun, S, and two planets, P, and P'.

---

[12] If this insistence on circles seems artificial—a silly way of dealing with planetary orbits—remember:

(i) that you have modern knowledge built into your own folklore,

(ii) that though this now seems to you an unreal model,

it is still fashionable as a method of analysis. Adding circle on circle in Greek astronomy corresponds to our use of a series of sines (projected circular motions) to analyze complex motions. Physicists today use such "Fourier analysis" in studying any repeating motion: analyzing musical sounds, predicting tides in a port, expressing atomic behavior. *Any* repeating motion however complex can be expressed as the resultant of simple harmonic components. Each circular motion in a Ptolemaic scheme provides two such components, one up-and-down, one to-and-fro. The concentric spheres of Eudoxus could be regarded as a similar analysis, but in a more complex form. Either scheme can succeed in expressing planetary motion to any accuracy desired, if it is allowed to use enough components.

CHAPTER 4 · AWAKENING QUESTIONS

[The Angel Raphael discusses the alternative views with Adam:]

> Hereafter, when they come to model Heav'n
> And calculate the Stars, how they will wield
> The mighty frame, how build, unbuild, contrive
> To save appearances, how gird the Sphere
> With Centric and Eccentric scribbl'd o'er
> Cycle and Epicycle, Orb in Orb:
>
> . . . .
>
> . . . What if the Sun
> Be Center to the World, and other Stars
> By his attractive virtue and their own
> Incited, dance about him various rounds?
> Their wandering course now high, now low, then hid,
> Progressive, retrograde, or standing still,
> In six thou seest, and what if seventh to these
> The planet Earth, so stedfast though she seem,
> Insensibly three different Motions move? . . .

—John Milton, *Paradise Lost*, Book VIII (1667)

Such was the Ptolemaic picture of the heavens—a complicated, clumsy system but workable and successful. To your mind and mine, it may seem unreal or unthinkable, but to Ptolemy and to many after him it was the alternative that seemed unthinkable, a spinning Earth whizzing round the Sun in a vast orbit. *Would not objects be thrown off a spinning Earth or left behind by a moving Earth?* And would not the nearer of the fixed stars change their apparent positions when the Earth moved right across the diameter of its orbit in half a year? Mankind's muddled ideas about motion had to wait for Galileo's teaching and Newton's clear thinking before the fog of the first objection cleared away. The second objection would still be serious if the stars did show absolutely no parallax shifts. We now know there are shifts, but they are too small to be observed without *very* delicate instruments. The first successful observation was made in 1832. These scientific objections to the Sun-in-center alternative were overshadowed by one that seemed to arise naturally from man's sense of his own importance. The Earth on which We live must be the center of the Universe—other things must revolve round Us.[1] This view, suggested by observation, gained easy support from simple self-centered thinking and from humanist teaching. It was firmly placed in people's beliefs. So we must not be surprised to find that Ptolemy's picture, with Man's world at the center, held the field in man's beliefs right on through the Dark Ages until the Renaissance brought questionings with awakening flexibility of mind.

The other view, with the Sun as center and a spinning Earth travelling round it, was suggested by some Greek astronomers; also discussed by an occasional philosopher or churchman in the 12th to 15th centuries, but the suggestions were put

---

[1] Hesitate before you condemn this arrogance unsympathetically. You, too, have passed through just such attitudes. Healthy children today start with much this kind of view: "I am the important person, almost the only person, and the world is arranged around Me." The process of growing up socially involves the broadening of this view, perhaps in stages such as "ME"; "Me and Mother"; "Me and family"; "Me and you"; "Me and other people"; "Me and my country"; and, finally, for a few, "me and the world." Most people fail to advance through all these stages. A man's success in social adjustment and wisdom would seem to depend on how far he progresses along this series. Many fail to advance beyond the first few stages, and though they may be very successful in material matters we are unlikely to find much to admire in them spiritually or intellectually. Some of the world's dictators got stuck at the "ME" stage, though others got as far as "Me and Mother" and regressed. At the other extreme, the rare souls who think and feel in terms of "me and the world" are the great philosophers and prophets.

apologetically as unreal theories, and gained little acceptance. For a thousand years the Ptolemaic picture was believed and hardly questioned. People in Europe had little interest in science, except as a basis for wordy discussions, arguing from authority instead of inferring principles from experiment. The Church of Rome was responsible for what teaching there was, and it treated science with the same dogmatic authority that maintained its religious structure. Any dispute with its teachings, even a simple appeal to experiment and observation, would bring disturbing questions to threaten its authority. Such disturbances were not welcome in an age when simple men were directed and taught in all matters by the Church and nobles and kings ruled by the authority of the Church.

In the thousand years between the Greek astronomers and the awakening of scientific experimenting there must have been questioning scientists, but their work remains forgotten. After centuries of "dark ages," came glimmerings of the new light. Roger Bacon, an English monk (~1250), almost cried aloud, "Experiment, experiment." He was an honest but intemperate critic, attacking churchmen and other thinkers in his insistence on the need to experiment and to collect real knowledge instead of poring over bad Latin translations. In his books he attacked ignorance and prejudice and made wise suggestions for gathering more knowledge. One writer pictures him shouting to mankind, "Cease to be ruled by dogmas and authorities; *look at the world!*"[2] His tactless manner brought him into conflict with his brother friars and with some Church authorities as he discussed and wrote. His teaching and books, and those of others like him, were probably suppressed and certainly forgotten for a long time—he was centuries ahead of his time. (The later Bacon, Francis, is credited with formulating the new scientific attitude, but some modern critics doubt whether his contribution was very helpful.)

Two hundred years later Leonardo da Vinci (~1480) thought and experimented and wrote and drew as a scientist as well as an artist. In mechanics he began to sort out ideas of force and mass and motion; and he formulated scientific ideas and sketched skillful models. His famous notebooks are a storehouse of mechanical inventions and include some of the finest drawings in the history of art. In making these notebooks, he was both historian and prophet, recording interesting ideas from others and ingenious schemes he had thought of. A new approach was starting which would have warmed Roger Bacon's heart.

Meanwhile, astronomical records had grown, with observations from Arab astronomers and others. The needs of medicine and navigation kept scientific teaching alive and ultimately impressed scientific growth on the Renaissance. Alphonso X of Castille (~1260), ordered his school of navigators to construct new tables to predict the motions of the heavenly bodies. These tables were collected, printed some 200 years later, and used for a further 100 years. Alphonso is rumored to have said, when he first had the complicated Ptolemaic system explained to him, that if he had been consulted at the Creation, he would have made the world on a simpler and better plan.

Others added measurements and in turn made mathematical refinements of the Ptolemaic machinery, but even in the intellectual awakening of the early Renaissance the alternative idea, Sun-at-center, was not put forward seriously until Copernicus wrote his great book. Then, running right through the Renaissance and up to the present day, came the great series of scientists who developed the science of mechanics from the foggy views of the dark ages to its present precise and powerful form, using the solar system (and, later, atomic systems) as a vast laboratory with frictionless apparatus. We are concerned not only with the physics they developed but also with the interaction between their work and the life and thought of other men. Therefore, we shall give some account of their lives as well as their work. First we shall give a set of short notes showing how their main contributions were related.

SHORT NOTES

[In these notes, as in earlier ones, the dates given are not the date the man was born or died but "average" dates usually showing when he was about 40.]

Nicolaus Copernicus, a Polish monk (~1510), suggested that the Sun-at-center (heliocentric) picture of the planetary system would be simpler. He wrote a great book setting forth the details of this system, showing calculations of its size, etc., and predicting tests. After his death this view spread, though it was not universally accepted until much later.

Tycho Brahe, a Danish nobleman (~1580) who, fired with curiosity about the planets, became a brilliant observer, a genius at devising and using

---

[2] H. G. Wells, *The Outline of History* (London, 1923).

precise instruments. He built the first great observatory. He knew of Copernicus' suggestion but did not wholly accept it, and was not much concerned with theory. He constructed far more accurate planetary tables than any before him, and left his pupil Kepler to complete their publication.

Johannes Kepler, a German (~1610), was a powerful mathematician with a gift for subtle speculation and a great belief that there are simple underlying rules in nature. Using Tycho's observations, he extracted three general rules for the motion of the planets. He could not find any underlying explanation of these rules.

Meanwhile, *Galileo Galilei* (~ 1610), was experimenting and thinking and teaching new scientific knowledge of mechanics and astronomy. To the dismay of classical philosophers, and to his own danger, he preached the need to abide by experiment. With the newly-invented telescope he gained evidence supporting Copernicus' picture which he advocated violently until the Church stopped him.

René Descartes, French philosopher (~1640), described a world system deduced from general principles which he felt were implanted by God. He opposed the idea of a vacuum and filled space with whirling vortices to carry the planets. His greatest contribution to science was the invention of co-ordinate geometry: the use of *x-y* graphs to link algebra and geometry. This enabled calculus to develop.

GREAT SCIENTIFIC SOCIETIES were formed for the exchange of knowledge, and experimental science throve publicly. (1600- . . . , 1700- . . . now.)

Isaac Newton (~1680), gathered up the results of Galileo's experimenting and the work and thinking of others into clearly-worded "laws" summing up the experimental facts concerning mass, motion, and force, with the help of clear ideas and definitions. He extended the force of gravity to universal inverse-square-law gravitation, showing that this would account for the moon's motion, for Kepler's three planetary laws, for the tides, etc., thus building a great deductive theory. In the course of this he needed calculus as a mathematical tool, and invented it. He experimented and speculated in other branches of physics, too, particularly optics.

In terms of his summary of mechanics, Newton showed that universal gravitation would "explain" the whole of the behavior of the Moon and planets as described by the Copernicus-Kepler system. In the next two centuries further consequences were worked out by other mathematicians and physicists, including the French mathematicians, *Joseph Louis Lagrange* and *Pierre Simon de Laplace*; and a new planet was discovered by its minute gravitational effects on the known ones.

Early in this century, *Albert Einstein* suggested modifications and reinterpretation of the laws of mechanics. These do not destroy Newton's work, but enable us to account for such things as a small unexplained motion of the planet Mercury, and to deal successfully with very rapid atomic motion. In addition to such modification of the "working rules" of mechanics, the great value of Relativity lies in the light it throws on the relation between experiment and theory, ruling out unobservable things from even the speculation of wise scientists.

CHAPTER 5 · NICOLAUS COPERNICUS (1473-1543)

Beauty is truth, truth beauty,—that is all
Ye know on Earth, and all ye need to know.
—John Keats

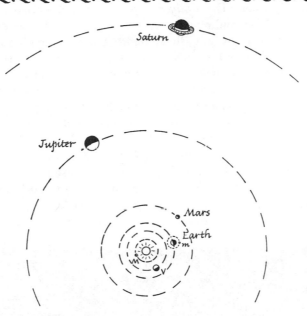

Fig. 16-1A. Copernicus' Planetary System

Nicolaus Copernicus was born in Polish Prussia. He seems to have lived a quiet, uneventful life. He was pious, capable, not brilliant, but inspired by a love of truth. He had the clear vision and the courage to challenge traditional authority, but he had no delight in entering into conflict with it.

Copernicus was brought up by his uncle, who was Bishop and ruler, practically prince, of the district. His uncle intended him to serve the Church and sent him to school and university near home to study the great classics. He also studied some astronomy and learned to use the clumsy astronomical instruments of the time. Then he travelled to Italy, learned Greek, and studied Church Law, in which he received the degree of Doctor. He also continued to study astronomy, and was now able to read the original Greek texts. After a few years, when he was 26, he visited Rome, and while there gave a course of lectures in mathematics (?=astronomy). Meanwhile his uncle had named him canon in the cathedral but allowed him to spend two more years in Italy to study medicine.

At last, in his early thirties, Copernicus returned to his uncle's cathedral near home. There he spent the rest of his life, as a scholarly monk, dividing his time between church duties, account keeping, occasional medical consultations, and meditation on the System of the World.

For his meditations he liked solitude, and he seems to have made few friends, though his reputation as a scholar drew several students to him. He had no use for the long wrangling arguments that were fashionable; yet, when asked to help a government committee which was trying to simplify the coinage, he accepted willingly and presented a clear capable report, which the senate adopted.

Copernicus was impressed by the variety and disagreement of opinions on planetary motions. The Ptolemaic system, with its artificial equants, seemed to him too clumsy to be God's best choice. He believed that the planetary system, spheres and all, was a divine creation; but he believed God's arrangement would be a simple one, all the more splendid for great simplicity. He collected together observations of the planets in more reliable tables than had so far been available; and in thinking about the planetary motions he was struck by the simplicity that would come from changing to a Sun-

56

in-center picture. He made an intuitive guess that the Earth is a planet like the rest—an extraordinary shift of view. Then he pictured all the planets moving in circular orbits around a fixed Sun. He made the Earth travel once around the Sun in a year, spinning once in 24 hours as it goes. The "fixed stars" and the Sun could then remain at rest in the sky.

This scheme replaced Ptolemy's epicycles and equants with simpler circular motions. The *daily* motion of the stars, carrying Sun, Moon, and planets as well, could obviously be replaced by a daily spinning Earth. That alternative had often been discussed, but had been turned down because the critics did not understand the mechanics of motion. (They claimed that there would be a howling wind of air left behind, and that the ground would outstrip a stone dropped from a high tower. On the other hand, the stars etc. could well be carried around by Ptolemy's spheres because spheres and rotations were natural in the heavenly region.)

The slower, irregular motions of Sun and planets through the star pattern could be simplified to circular motions around the Sun. This was Copernicus' main contribution: to stop the Sun and place it at the center of the planetary system. Then the Sun's yearly motion around the ecliptic was only an apparent one due to the Earth's yearly motion around the Sun. And the complex epicycloid of a planet was simply a compound of the planet's own motion around a circle and the Earth's yearly motion. (On this view, the epicycloid picture is making us pay for ignoring the Earth's motion.) This tempting idea of a Sun-in-center scheme had been thought of before, but not strongly supported. Copernicus, searching early records for such ideas, had both the clear mind and the store of data to develop it.

His detailed explanation of a planet's epicycloid ran like this. Suppose the Earth travels around a circular orbit and Jupiter more slowly around a bigger orbit, both with the Sun at the center, as in Fig. 16-2. The fixed stars must be *much* farther away, because no parallax-shifts are observed. Then in marking the position of Jupiter among the fixed stars we look along a sightline running from Earth to Jupiter and on, far beyond, to the pattern of the

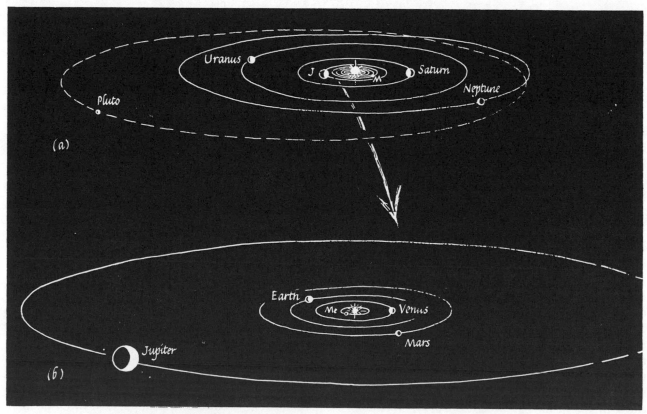

Fig. 16-1B. The Copernican System (with later additions) seen in perspective
(a) The whole system.   (b) Inner region of (a) magnified several times.
The orbits are almost circular, with the Sun only a little off center.
The arrangement of orbits is shown here, but the sizes are *not to scale*.
The planets themselves are *much* smaller in proportion than these sketches show—on the scale
used here for orbits the planets would be invisible dots.

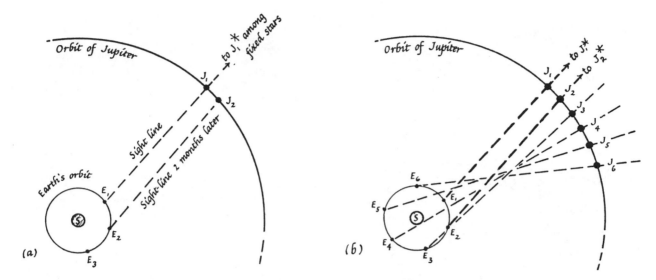

Orbit of Jupiter

FIG. 16-2. COPERNICUS' EXPLANATION OF PLANETARY EPICYCLOIDS
The lines E_1J_1, E_2J_2, etc., are sight-lines from positions of the Earth every two months through Jupiter's position towards the stars.
(a) Two stages sketched   (b) More stages sketched.

stars. As the Earth sweeps round and round its orbit and Jupiter crawls more slowly, this sightline wags to and fro as it goes around, marking an epicycloid among the stars. When the Earth is at E_1, Jupiter is at J_1, and an observer looking along the sightline E_1J_1 sees Jupiter among the stars at J_1^*. As the Earth travels from E_1 to E_2 to E_3, E_4, E_5, E_6, &c., Jupiter travels steadily but slowly forward from J_1 to J_2 to J_3, J_4, J_5, J_6, &c. Then the observer on E sees J^* in directions that swing mostly forwards but sometimes backwards. To see this, look at Fig. 16-3, which shows Fig. 16-2 condensed to a small scale with the sightlines continued out to a remote background of stars.

Copernicus accounted for the epicycloids of Mars, Jupiter, and Saturn by making them move around large circular orbits outside the Earth's orbit. He made Venus and Mercury move around smaller orbits, nearer the Sun than the Earth's. This accounted for their observed behavior—keeping close to the Sun and swinging to and fro each side of it. Thus the same scheme served for both the "inner" planets and the "outer" ones.

Copernicus did not just offer an alternative that looked simpler; he extracted new information from his scheme: the order and sizes of the planetary orbits, a remarkable advance contributed by theory. In the Ptolemaic scheme the main circles could be chosen with any sizes—it did not even matter which planet was put outermost. In fact, Ptolemy was just drawing patterns with a mathematical machine, to fit the observations, on the celestial sphere. In the

Sun-in-center scheme, the orbits must be in a definite order and must have definite proportions. From the planets' apparent motions in the sky it was obvious to Copernicus whose orbits were largest and whose least.[1] The order must be (as in Fig. 16-1):

SUN, stationary at the center
Mercury, nearest the Sun
Venus
Earth, with the Moon travelling around it
Mars
Jupiter
Saturn, farthest of the planets then known.

Treating the orbits as simple circles, Copernicus calculated their relative radii from available observations; he could thus plot a fairly accurate scale map of the system. To obtain the actual radii from these relative values, he needed an absolute measurement of any one of them, say the distance from Sun to Earth. This was known only roughly,[2] so the *scale* of his complete picture was unreliable.

[1] Try this theoretical discussion, with Figs. 16-2 and 16-3. Suppose that Jupiter and the Earth each keep their present times of revolution (our year, and Jupiter's "planetary year"), but that the radius of the Earth's orbit is changed. What would happen to the shape of the loops of Jupiter's apparent path, among the fixed stars,
  (a) if the Earth's orbit changed to a *very small* radius?
  (b) if the Earth's orbit changed to be nearly as big as Jupiter's?
[2] In fact he used a Greek estimate that was 20 times too small.

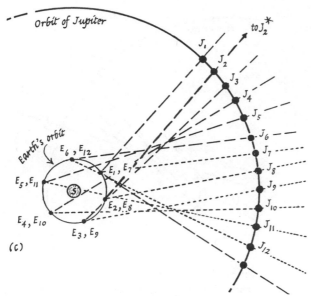

(c) Many stages sketched. The sight-line EJ wags up and down in a complicated way.

Estimating Orbits

To see how he calculated relative radii, suppose you are attacking the problem for an inner planet, say Venus. Venus, nearer the Sun than the Earth, travels in a small orbit round the Sun. This circle is seen practically edge-on from the Earth; so Venus seems to swing to-and-fro in front of the Sun or behind it, travelling only a small way each side of the Sun before it turns back. Thus it is seen only near the Sun as a morning or evening "star." When Venus seems farthest to one side of the Sun, just about to turn back, it must be at a point such as C lying on a tangent from the Earth to its orbit (Fig. 16-4). In positions A, B, D, . . . etc., it would seem nearer the Sun. This tangent is perpendicular to the radius, SC, of the orbit (by geometry of circle). So the triangle ECS has a right angle at C and an angle at E that can be measured by sighting from the Earth. Given these, you can draw a scale model of this triangle (i.e., a similar triangle), and by measuring that you can find the proportion between SC and SE which are the radii of the orbits of Venus and the Earth. To measure the required angle at E, you must find the angular distance between Venus and the Sun at the moment when Venus seems farthest from the Sun. In trying to make a direct measurement you might be prevented by the glare of the Sun, but you can wait till the Sun has set, calculating where it will have got to, and then observe Venus day after day till the separation is greatest. So you may imagine yourself measuring the angle with two jointed sticks with peep-

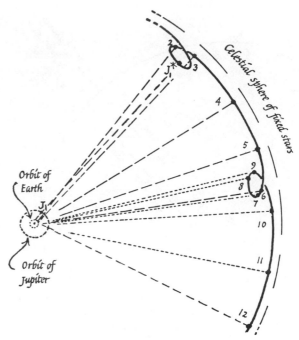

FIG. 16-3. COPERNICUS' EXPLANATION
The apparent positions of Jupiter in the background of fixed stars. This shows FIG. 16-2c redrawn on a much more condensed scale with the sight-lines from Earth to Jupiter continued on out to the stars. (E.g. the line to J_1° here is continuation of EJ_1.) The specimen sight-lines are drawn parallel to the corresponding ones in FIG. 16-2c.

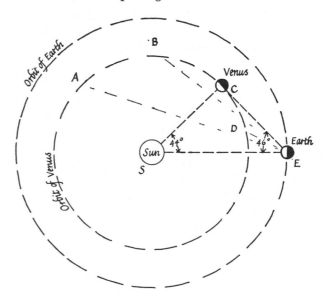

FIG. 16-4. ESTIMATING RELATIVE RADII OF ORBITS
Venus is shown farthest from the sun.

holes, though the real method must be slightly less direct. Actual observations show that this angle is about 46°. Drawing and measuring a triangle with angles 46°, 90°, and 44° will show you that the fraction (medium side) / (longest side) is about $^{72}/_{100}$. This tells you that the orbit-radii for Venus and Earth are in the proportion 72:100. You need not draw the triangle, if you have trig. tables, as Copernicus had. The fraction you want (medium side)/(longest side) is sin 46°, and, from tables, this is 0.72. Copernicus had measurements which gave him this angle, and he performed this calculation for Venus and Mercury. For the outer planets the argument and the geometry are rather more complicated, but Copernicus calculated the relative sizes of their orbits in much the same way. Thus, he could draw a scale map of the orbits and place the planets correctly in them at some chosen starting time. To predict their positions at other times he needed to know each planet's "year," the time it takes to travel round its orbit. These "years," or times of revolution, he found from recorded observations. Essentially, he found how long the planet took to get back to the same place among the stars.

Using recorded data, Copernicus placed the planets on his scale map and predicted their positions at other times, past and future. He could check the past ones, and thus test his "picture," or "theory" as we should now call it. These tests were encouraging, but there were some disagreements which led, through long careful calculations, to modifications of the simple picture.

Copernicus gave other points in support of his theory:

(i) Mars is much brighter (seeming larger) in some seasons, obviously because it is nearer the Earth then. On the Ptolemaic system its slightly eccentric orbit round the Earth could not possibly provide big enough changes of distance. But on the Copernican scheme the distance ranges from the sum of the orbit radii to the difference. In fact Mars *is* brightest when that distance is least, at the seasons when Mars and the Earth are on the same side of the Sun—Mars in "opposition" to the Sun, overhead at midnight.

(ii) Just when an outer planet makes the *reverse* part of a loop it is exactly in opposition to the Sun. Ptolemy could give no reason for this. It is obvious from the Copernican geometry (study Fig. 16-2).

(iii) If Venus and Mercury are nearer the Sun than the Earth is and travel round the Sun in small orbits, then when we look at them we should see only part of them brightly lit, the side facing the Sun, not the whole planet (see Fig. 16-5). Thus these two planets should show stages or "phases," like the Moon as it changes from new Moon to half Moon to full Moon and so on.[3] With the large

[3] On a pure Ptolemaic scheme, Venus would also show phases but not the whole range from crescent to "half moon" to full.

planet Venus, these stages should, the critics thought, be visible. As they were not observed, the critics claimed the Sun-in-center idea was wrong. There is a story, almost certainly mythical, that Copernicus replied: "If the sense of sight could ever be made sufficiently powerful, we should see phases in Mercury and Venus." Within a century, Galileo's telescope showed the phases of Venus.

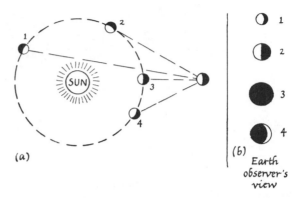

FIG. 16-5. PHASES OF VENUS, AS SEEN FROM THE EARTH

As a crowning virtue of simplicity, Copernicus gave a new interpretation of the precession of the equinoxes. Precession, as discovered by the Greeks, was described as the whole star system (and the Sun) crawling slowly around the axis of the *ecliptic*, while the Earth and its equator plane and N-S axis stayed still. Copernicus reversed the description, saying the Sun and its ecliptic plane stay fixed; that is, the plane of the Earth's orbit stays fixed. And the Earth's equator-plane (and celestial equator) swings slowly around, always tilted 23½° to the ecliptic. Then Copernicus could describe precession simply: the Earth's spin-axis has a slow conical movement; carrying the equator-plane, it gyrates around a cone of angle 23½° in 26,000 years. Though Copernicus gave this clear picture of what happens in the precession of the equinoxes, he had no idea what "caused" it. That problem had to wait for Newton, who showed that, like so many astronomical phenomena, it is a result of universal gravitation.

Though the simplicity delighted him, Copernicus presently found that steady motion in simple circular orbits would not fit the facts. He had to make the orbits eccentric, and even add little epicycles. In doing this he spoiled the simplicity somewhat, and perhaps made it harder to guess the underlying simple rules which Kepler found later; but, like Ptolemy, he was insistent on making his machinery fit observations accurately. In this respect both sys-

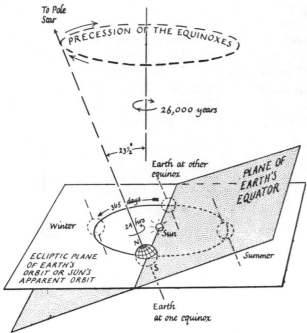

To Pole Star

PRECESSION OF THE EQUINOXES

26,000 years

23½°

Earth at other equinox

PLANE OF EARTH'S EQUATOR

365 days

24 hrs

Sun

Winter

N

S

Summer

ECLIPTIC PLANE OF EARTH'S ORBIT OR SUN'S APPARENT ORBIT

Earth at one equinox

FIG. 16-6. SKETCH SHOWING MOTION CALLED THE PRECESSION OF THE EQUINOXES

tems were good descriptions of the observed motions, and we ought not to call either "wrong."

Copernicus spent twenty years or more constructing and perfecting his scheme. During this time he became well known among mathematicians and astronomers, and some came to talk and study with him and carry away his powerful idea. He sent a

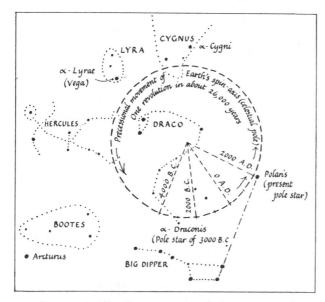

CYGNUS
α-Cygni

LYRA

α-Lyrae (Vega)

Precessional movement of Earth's spin-axis (celestial pole). One revolution in about 26,000 years

HERCULES

DRACO

2000 A.D.

4000 B.C.

2000 B.C.

O A.D.

Polaris (present pole star)

BOOTES

α-Draconis (Pole star of 3000 B.C.)

Arcturus

BIG DIPPER

FIG. 16-7. THE PRECESSION OF THE EQUINOXES
Sketch of a large patch of Northern sky (about 90° by 90°), showing the slow movement of the celestial North Pole among the stars. The point where the Earth's spin-axis cuts the pattern of the stars moves slowly around a roughly circular path making one revolution in about 26,000 years. (After Sir Robert Ball.)

small outline of his scheme to friends. Yet he had no wish for fame. Many urged him to publish his work: his star maps, his tables of observations, and his great scheme of the Solar System with its full defense and all the details he had worked out. Even after a friend had published a preliminary review of his scheme, he delayed for many years because he wanted to correct and improve it before he offered such a revolutionary change of view to a conservative public. He had no fear of conflict with the Church of which he was a well-accepted member; and the friend who offered to pay for the publishing was himself a Cardinal. But he knew that such a book would arouse opposition, and he feared ridicule. Great clearness, critical reasoning, and organized data would be needed to convince a prejudiced world. It was a tremendous thing to try to overthrow the Ptolemaic system, founded and perfected by the great men of the past and made almost sacred by tradition and practical use. For a long time the Ptolemaic scheme had been failing in accuracy—even the calendar had accumulated a big error. Yet no one doubted its essential rightness. Astronomers merely tried to correct its radii and shift its equants for better agreement. Copernicus wanted to be very sure of his ground. He could afford to wait, for he knew that truth is very strong.

At last he was persuaded, and he wrote a great book which was set up in type and published at the very end of his life. Its title was *De Revolutionibus Orbium Cœlestium*, "On the Revolutions of the Heavenly Spheres." He dedicated it to the Pope, saying in his dedication,

"I can easily conceive, most Holy Father, that as soon as some people learn that in this book . . . I ascribe certain motions to the Earth, they will cry out at once that I and my theory should be rejected. For I am not so much in love with my conclusions as not to weigh what others will think about them, and although I know that the meditations of a philosopher are far removed from the judgment of the laity, because his endeavor is to seek out the truth in all things, so far as this is permitted by God to the human reason, I still believe that one must avoid theories altogether foreign to orthodoxy. Accordingly, when I consider in my own mind how absurd a performance it must seem to those who know that the judgment of many centuries has approved the view that the Earth remains fixed as center in the midst of the heavens, if I should, on the contrary, assert that the Earth moves; I was for a long time at a loss to know whether I should publish the commentaries which I have written in proof of its motion."

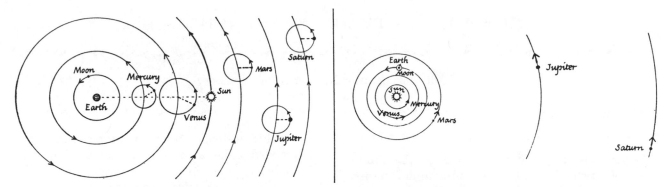

FIG. 16-8a. PTOLEMAIC SYSTEM,
sketched without eccentricity or equants.
Order and proportions of orbits not determinate.
Epicycle radii not "to scale."

FIG. 16-8b. COPERNICAN SYSTEM,
sketched without eccentricity or minor epicycles.
Orbit proportions, which *are* determinate, are
roughly to scale. (Moon's orbit out of scale.)

FIG. 16-8. COMPARISON OF SIMPLE PTOLEMAIC
SCHEME AND SIMPLE COPERNICAN SCHEME

He further says,

"If there be some babblers, who, though ignorant of all mathematics, take upon them to judge of these things, and dare to blame and cavil at my work, because of some passage of Scripture which they have wrested to their own purpose, I regard them not, and will not scruple to hold their judgment in contempt."

In the book he gave his star tables and planetary observations as well as his great exposition of the new view of the Solar System.

The book was set in type and printed far away, in Nuremberg. It starts with a general description of the scheme and its advantages.[4] Then chapters expound the necessary trigonometry, and a section develops the rules of spherical astronomy. Then the Sun's "motion"—or rather the Earth's—is discussed in full, with the explanation of precession of the equinoxes. The Moon's motion is then discussed, and the last two sections deal with the motions of the planets in full detail. In the latter, he included discussions of measurements of the distances of Sun and Moon and the sizes of planetary orbits.

It was a great book indeed, destined to have far-reaching effects. Copernicus never read it in its final printed form. While he was waiting for its publication, an old man of 70, he was taken very ill, partly paralyzed. On May 23, 1543 the first printed copy was sent him, so he saw it and touched it. That night he died peacefully.

[4] A friend added a timid preface to say this was only a theory—far from that, Copernicus believed his scheme was *true*.

CHAPTER 6 · TYCHO BRAHE (1546-1601)

"The fault, dear Brutus, lies not in our stars. . . ."

The Copernican Revolution

Copernicus struggled to escape from the atmosphere of Aristotelian dogmatism and argumentation in which he had been brought up. He loved truth, and he succeeded in producing a clearer, better picture of the heavens. Though insistent, he was calm and peaceful and considerate, and he persuaded many to his view. After his death, knowledge of his system spread. His tables were checked, corrected, and printed. His estimate of the length of the year (365 days, 5 hours, 55 minutes, 58 seconds) was used by a later Pope to reform the calendar, which was by then days out of gear with the seasons. When Copernicus himself had been consulted on calendar reform some eighty years earlier, he had demurred because he wanted first to clear up his own ideas about the system of the world. Here is a tribute to his final work: the reformed calendar, which we still use, has its scheme of leap years so well arranged that it will remain true to the Sun's seasonal calendar within *one day* for the next 3,000 years!

But as time went on others taught the new system less quietly and less tactfully, and its real effect began to be felt. It upset philosophers, disturbed ordinary people, started scientists towards new thinking, and brought the official Church into open opposition. Within a century of his death, Copernicus' work, originally dedicated to the Pope, became the center of one of the most violent intellectual controversies that the world has known. How could it do this? Because it upset knowledge that was taken for granted and seemed obviously right; and because it upset the intellectual outlook of the times. Copernicus was attacking—though he did not realize he was—a great interwoven structure of thought and belief. In those days scholarly knowledge was not divided into separate fields of study such as physical science, biology, physiology, psychology, sociology, languages, arts, and philosophy. We who are taught in organized classes and specialized courses can hardly picture the confused but strongly knit *general* intellectual outlook of the medieval intellectual world. The medieval scholars who met and discussed and taught were masters of all fields, anxious to maintain a scheme that gave all knowledge the same assured standing. They tied all studies together in a unified system.

The natural world was regarded as made up of four basic elements: earth, air, fire, and water. Liquids flowed because they were mostly water; solids were dense and strong because they were mostly earth; and so on. Four corresponding "humours" ruled men's behavior—according to which of the four is dominant in him, a man is angry, sad, calm, or strong in his temperament. The planets, marked out among the stars by their wandering, were given qualities that linked them both to man and to the four physical elements. They were linked to human temperament and fate, and they were linked to the metals whose properties depend on their proportions of earth and fire. (For example, Mars, the reddish planet, might represent the god of war, angry temperament, fiery metal.) The planets and stars in turn linked man to divine providence. The celestial system of spheres formed a vast ideal model for man himself. And beyond the outermost sphere of stars was heaven—placed on top of the astronomical world in the proudest place of the celestial scheme. Thus an attack on the planetary system threatened to change all teaching about nature and man, and even man's relation to God.

But did Copernicus attack the planetary system with its moving spheres? He only moved the center from the Earth to the Sun. To see why that upset the whole system of thought, first see why the spheres were taken for granted.

Professor Herbert Dingle gives one historian's view as follows:

"They were not seen or directly observed in any way; why, then, were they believed to be there? If you imagine the Earth to be at rest and watch the sky for a few hours, you will have no difficulty in answering this question. You will see a host of stars, all moving in circles round a single axis at precisely the same rate of about one revolution a day. You cannot then believe that each one moves independently of the others, and that their motions just happen to have this relation to one another. No one but a lunatic would doubt that he was looking at the revolution of a single sphere with all the stars attached to it. And if the stars had a sphere, then the Sun, Moon and planets, whose movements were almost the same, would also be moved by the same kind of mechanism. The existence of the spheres having thus been established and accepted implicitly for century after century, men no longer thought of the reasons which demanded them, but took them for granted as facts of experience; and Copernicus himself, despite his long years of meditation on the fundamental problems of astronomy, never dreamed of doubting their existence."[1]

But presently the full meaning of Copernicus' work was realized. A spinning Earth would remove any need for a spinning bowl to carry the stars. The stars needed no bowl: they could just hang in the sky at *any* great distance, and if at *any* perhaps at *many different* distances. That pushed the stars back into the remoter region of Heaven. That was a terrifying change—where was the Heaven of theology now to be located, the Home of God and the abode of departed souls? A century after Copernicus, Giordano Bruno suggested such a picture of infinite space, peopled with stars which were distant suns. He was burnt at the stake as a heretic for his unfortunate views.

Again, Copernicus pushed the Earth out of its ruling position as center of the universe, to join Mars and Jupiter, as a planet ruled by the Sun. That made the machinery of spheres seem less complex and less necessary; and it completely upset the picture of planets holding controlling influences over a central Earth and Man.

"We can hardly imagine today the effect of such a change on the minds of thoughtful men in the sixteenth century. . . . With the spheres went a localised heaven for the souls of the blessed, the distinction between celestial and earthly matter and motion became meaningless, the whole place of man in the cosmic scheme became uncertain.

". . . Copernicus himself . . . was thoroughly medieval in outlook, and had he been able to foresee what his work was to do we may well believe that he would have shrunk in the utmost horror from the responsibility which he would have felt. But what he did was to make it possible for the new scientific philosophy to emerge."[1]

[1] From *Copernicus and the Planets*, a talk by Herbert Dingle, Professor of the History and Philosophy of Science in the University of London. The rest of the discussion on this page and the previous one also draws on Professor Dingle's talk, published as Chapter III of a symposium of talks on the *History of Science* (Cohen and West, Ltd., London).

If this was an upheaval for scholars, what might it not do to the vast uneducated crowd once they understood it? Not only was the machinery of the spheres wrecked, but the general outlook on knowledge was threatened with a change, and so was men's faith in God's provision for their souls.

Copernicus' work reached the common people through its effect on astrology. For many centuries, superstitious people—almost everyone, from king to beggar—believed that the Sun, Moon, and planets influenced their fate. Casting a man's horoscope was a system of fortune-telling based on the positions of the planets at his birth. A horoscope by an accredited astrologer was trusted to help kings rule wisely (or safely) and to foretell any man's character and future. As long as the Earth was considered the center of the heavenly system, with planets wandering strangely around our important selves, it was natural to think that planetary motions occurred for our benefit and might even control our fate. When Copernicus moved the Sun into the center of the planetary system *and described the Earth as merely one of the planets*, he started the downfall of astrology from its position as counsellor of kings and comforter of common men to its present rank as amusing fiction. Yet civilized man is still gullible enough—insecure enough, or perhaps romantic enough—to keep the publishing of astrology books a thriving business.

Opposition to the Copernican system arose from the Church, probably for two reasons; first because the system contradicted the Church's teaching of astronomy and philosophy; and second because preaching the new system meant a questioning of authority—questioning, even defying, the authority and tradition of the Ptolemaic system. Rebellious people who like new ideas, and enjoy proving others wrong, seldom find themselves popular. In those days the Church was anxious to maintain its tremendous power and insisted that all people obey its authority strictly and believe its teachings completely. Anyone who questioned the Church's authority or disagreed with its teaching was, to his own knowledge, risking his life and, in the Church's view, risking his soul. There were a few such men at that time. They were the martyrs. Though a scientist preaching a new *theory* might seem harmless, one insisting that it was *true* would seem an annoying rebel. He who rebels in one field may rebel in other fields, too, so the provocative arguer in any field is apt to strike the Church (or the State—authority wherever it is) as a dangerous man politically. Within less than a century Copernicus' book,

De Revolutionibus Orbium Cœlestium, had been placed on the list of forbidden books by the Church. (It remained there for 200 years, and was quietly dropped about 1830.)

In this stormy time for Astronomy, came Tycho Brahe, the great observer whose amazingly accurate measurements formed the basis for Kepler's discoveries and in turn Newton's explanations. Galileo lived at the same time as Kepler, and bore some of the worst of the storm. A few centuries later, Copernican advocates seemed harmless. As Lodge[2] points out, this happens in age after age. In the early 1800's geologists met with violent condemnation as impious critics of the Bible story of Creation. Later in the century geology was safe, but theories of evolution were condemned and the teaching of them forbidden—perhaps this continues. *Every* age has one or more groups of intellectual rebels who are persecuted, condemned, or suppressed at the time but, to a later age, seem harmless. Some are cranks and others are wise prophets—pacing the growth of man and man's knowledge—men ahead of their time.

Tycho Brahe

Tycho Brahe was the eldest son of a noble Danish family—"as noble and ignorant as sixteen undisputed quarterings [of their coat of arms] could make them." Hunting and warfare were regarded as the natural aristocratic occupations: though books were fashionable, study was fit only for monks, and science was useless and savoured of witchcraft. Tycho would probably have been brought up to be a soldier had not his uncle, a man of more education, adopted him, his parents agreeing unwillingly when their second son was born. His uncle gave him a good education. Tycho began Latin at the age of 7; his parents objected, but his uncle urged this would help him with law. At 13 he went to the university to study philosophy and law, and there his interest was suddenly turned toward astronomy by a special event. An eclipse of the Sun took place. The astronomers had predicted it, and people turned out to watch for it with great excitement. Tycho watched and was amazed and delighted. When it

happened at the predicted time it awoke in him a great love for the marvelous that made him long to understand a science that could do such wonders. He continued to study law, but his heart was set on astronomy. He spent his pocket-money on astronomical tables and a Latin translation of Ptolemy's *Almagest* for serious study. After three years, his uncle sent him abroad with a tutor, to travel and to read law in German universities. There he continued to work at astronomy secretly. He sat up much of each night, observing stars with the help of a celestial globe as small as his fist. He bought instruments and books with all the money he could get from his tutor without revealing his purpose. He found the tables of planetary positions inaccurate. Ptolemaic tables and Copernican tables disagreed, and both differed from the facts. As a boy of 16, he realized— what the professional astronomers of Europe had missed—that a long series of precise observations was needed to establish astronomical theory. A few planetary observations made at random could not decide between one system for the heavens and another. Here was the start of his life work.

When he was 17 he observed another special event, a conjunction of Jupiter and Saturn. Two planets are in "conjunction" when they cross the same celestial longitude together, very close to each other. This strange "clashing" of two planets could be predicted with the help of tables such as Ptolemy's or Copernicus'. Such events were regarded by superstitious people as bringing good or bad luck. The enthusiastic young Tycho observed the conjunction and compared his observed time with the predictions of the planetary tables. He found Alphonso's revision of the Ptolemaic tables wrong by a month,[3] and Copernicus' tables wrong by several days. He decided then to devote his life to the making of better tables—and he succeeded with a vengeance. He became one of the most skillful observers the world has known. Neither his aristocratic birth nor his education had saved him, however, from being superstitious, full of belief in occult influences; and he believed that this conjunction foretold, and was responsible for, the great plague which soon after swept across Europe.

Tycho started observing with a simple instrument: a pair of jointed sticks like compasses, one leg pointed at a planet and the other at a fixed star. Then he measured the angular separation by plac-

[2] I owe a debt of inspiration and delight to Sir Oliver Lodge, who first showed me how the history of astronomy illuminates physics, in his *Pioneers of Science* (Macmillan, London, 1893). In the next few chapters, I have drawn on Sir Oliver's book for general ideas and in many places for words and phrases. I am grateful to him and to those earlier writers on whom he drew in turn. Modern historians of science have weeded out some mistaken information, and they quite rightly plead for a wider view with less hero worship; yet the book was the work of a physicist and a great man with vision.

[3] A month's error may seem large in predicting a meeting of planets; but those tables dated back, essentially, to Ptolemy fourteen centuries before. One month in 1400 years seems little enough, a great credit to the Ptolemaic system as representing the facts accurately, however clumsily.

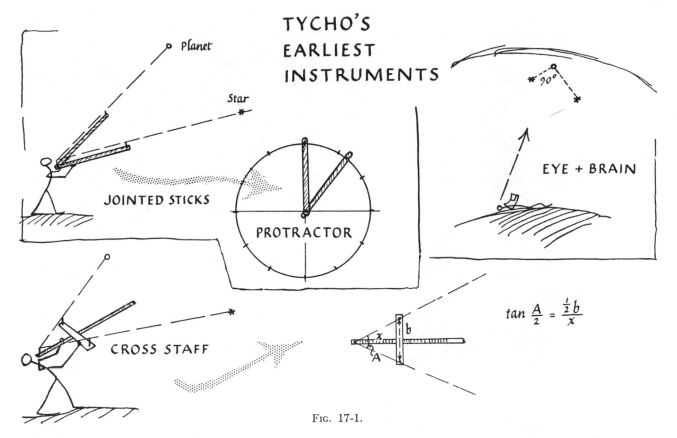

Planet

Star

JOINTED STICKS

PROTRACTOR

CROSS STAFF

EYE + BRAIN

$\tan \dfrac{A}{2} = \dfrac{\frac{1}{2}b}{x}$

FIG. 17-1.

ing the compasses on a graduated circle drawn on paper. He often used eyes + brain alone to mark a planet's position when it formed a 90° triangle with two stars that he knew. Soon he obtained a "cross staff," a graduated stick with a slider at right angles that carried sights at its ends. An observer looking through a peephole at the near end of the main stick could set the sights on two stars, and thus measure the angle between them. He found his instrument was not accurately graduated, so he made a careful table of corrections, showing the error at each part of the scale—a method of precision that he used all his life. This is the way of the best experimenters: for great precision they do not try to make an instrument "perfectly accurate," but they make it robust and sensitive, and then they calibrate it and record a trustworthy table of errors.

Presently he was called home by threats of war. His uncle died and the rest of his family did not welcome him. They blamed him for neglecting law, and they despised his interest in star-gazing. Disappointed, Tycho left Denmark for Germany to continue his studies. In his travels, he made friends with some rich amateur astronomers in Augsburg. He persuaded them that very precise measurements were needed, and they joined in constructing an enormous quadrant for observing altitude-angles by means of a plumb line and sighting holes fixed on the arm of a huge graduated circle. This wooden instrument was so big that it took twenty men to carry it to its place in a garden and set it up. Its circle had a radius of 19 feet. It had to be big for accurate measurements—there were no telescopes in those days, merely peep-holes for sighting. The quadrant was graduated in sixtieths of a degree. Tycho and his enthusiastic friends also had a huge sextant of radius about 7 feet. This was the beginning of his accurate planetary observations.

In his stay in Germany he met with a strange accident. His violent temper led him into a quarrel over mathematics, and that led to a duel which was fought with swords at seven o'clock one December night. In the poorly lit fight, part of Tycho's nose was cut off. However, he made himself a false nose (of metal or putty, probably painted metal). He is said to have carried around with him a small box of cement to stick the nose on again when it came off.

After four years in Germany, Tycho went home again, this time to be received well as an astronomer of growing fame. His aristocratic relatives thought more kindly of science and received him with admiration. When Tycho's father died, another uncle welcomed Tycho to his estate and gave him an

QUADRANS MAXIMUS QUALEM OLIM
PROPE AUGUSTAM VINDELICORUM EXSTRUXIMUS

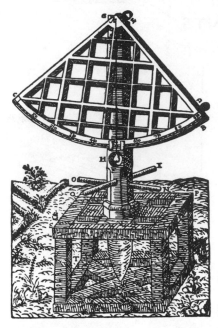

FIG. 17-2. TYCHO'S EARLY QUADRANT, built with his friends when he was a young man, travelling in Germany. (Radius of circle: about 19 feet.) Sighting the Sun or a planet through peepholes D, E, the observer could read its "altitude" by the plumb line AH on the scale graduated in sixtieths of a degree. This picture and the following ones are reprinted from *Tycho Brahe's Description of His Instruments and Scientific Work* by permission of The Royal Danish Academy of Sciences and Letters.

extra house as a laboratory for alchemy. Tycho's fascination with the marvelous had drawn him to alchemy. This was not a complete break with astronomy because the astrology of the time linked the planets closely with various metals and their properties. Alchemy had a useful side too: it gave Tycho knowledge of metals for instruments. Thereafter he often combined a little alchemy with his astronomical work and even concocted a universal medicine.

The New Star

The year after his return, a new star blazed up in the sky and was visible for many months. At first it was as bright as Venus, and could even be seen in daylight.[4] Tycho, amazed and delighted, observed it carefully with a large sextant, and found it was very far away, one of the fixed stars, "in the 8th sphere, previously thought unchangeable." After much careful watching and recording, Tycho published a report on it.

Tycho's fame was growing, and a group of young nobles asked him to give a course of lectures on astronomy. He refused at first, thinking this below the dignity of one of noble birth; but he was persuaded when he received a request from the King.

At this time, Tycho married a peasant girl—to the horror of most of his family—and thereafter seems to have modified some of his aristocratic prejudices.

The Great Observatory, Uraniborg

Finding his life as a noble interfered with astronomy, he embarked on another move to Germany; but King Frederik II of Denmark, understanding that Tycho's work would bring the country great honor, made him a magnificent offer. If Tycho would work in Denmark, he should have an island for his observatory, estates to provide for him, a good pension, and money to build the observatory. Tycho accepted with enthusiasm. Here at last was a chance to carry out his ambitions.

Tycho built and equipped the finest observatory ever made—at enormous cost.[5] He called it Uraniborg, The Castle of the Heavens. It was built on a hill on the isle of Hveen, surrounded by a square wall 250 feet long on each side, facing North, East, South, and West. In the main building there were magnificent living quarters, a laboratory, library, and four large observatories, with attic quarters for students and observers. There were shops for making instruments, a printing press, paper mill, even a prison for recalcitrant servants. Tycho made and installed a dozen huge instruments and as many smaller ones. These instruments were the best that Tycho could devise and get made—all constructed, graduated, and tested with superb skill and fanatical attention to accuracy. Some of them were graduated at intervals of $1/60$ degree, and could be read to a fraction of that.

In the library, Tycho installed the great celestial

---

[4] We now know that a new star, or *nova*, appears in the sky fairly frequently—some sudden condensation, or other change, heats a star to higher temperature. Much more rarely—averaging once in several centuries in our galaxy—

there is a far brighter outburst, a *supernova*. Hipparchus probably saw one, and Tycho's new star was one. A recent speculative theory suggests that the appearance of a supernova involves the radioactive element californium. Tycho's careful comparisons of the brightness of his star with standard stars as it died down fit well with the "half life" of californium—a fantastic modern use of his careful work.

[5] Some years later, Tycho stated the total cost. His estimate was equivalent to about 17,000 English pounds at that time. Translated in terms of cost-of-living this would be at least $200,000 today. Translated in terms of luxury and equipment, Uraniborg would be a multimillion dollar observatory.

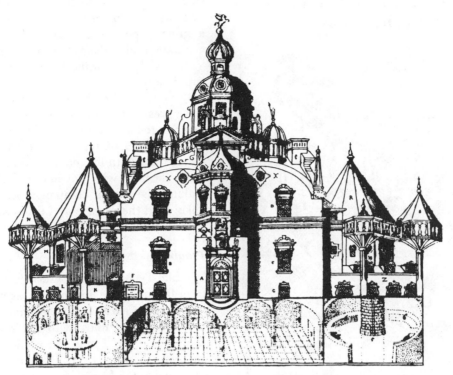

FIG. 17-3. URANIBORG. Design of the main building, built about 1580.

globe he had ordered in Augsburg some years before. It was covered with polished brass, an accurate sphere as high as a man. As the work of the observatory proceeded, star positions were engraved on it. Making and marking it took 25 years in all.

In Tycho's study, a quadrant was built on the wall itself, a huge arc with movable sights for observing stars as they passed a peg in a hole in the wall opposite. This was one of his most important instruments, and Tycho had the empty wall space inside the arc decorated with a picture showing himself and his laboratory, library, and observatories. Fig. 17-5 is an engraving of this mural, with observers in front using the quadrant and the primitive unreliable clocks of Tycho's day. (Tycho said the portrait was a good likeness.)

It was a gorgeous temple of science, and Tycho worked in it for twenty years, measuring and recording with astounding precision. Students came from far and wide to work as observers, recorders, and computers. This was Tycho's great work, to make continuous accurate records of the positions of Sun, Moon, and planets. Then he proposed to make a theory for them. At first he did not concern himself much with theory—though he insisted that without *some* theory an astronomer could not proceed with his work. Later in life he put forward a useful compromise that acted as a stepping stone for thinkers

GLOBUS MAGNUS ORICHALCICUS

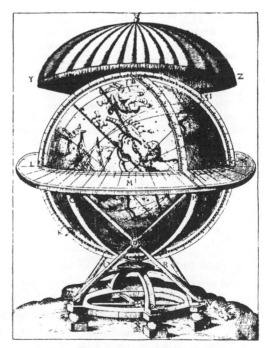

FIG. 17-4. TYCHO'S GREAT GLOBE

Tycho had this globe made very carefully, at great expense, so that he could mark his measurements of star positions on its polished brass surface. He ordered it before he started Uraniborg, had it brought and installed there, and took it with him when he moved to Prague.

QVADRANS MVRALIS
SIVE TICHONICVS

FIG. 17-5. TYCHO'S MURAL QUADRANT
The huge brass arc was firmly fixed in a western wall,
with its center at an open window in a southern wall.
The empty wall above the arc was decorated by a huge
painting showing: Tycho observing; students calculating;
Tycho's globe, books, dog; and some of Uraniborg's
main instruments. An observer sighted the star (or Sun)
by pinholes at F and a marker in the window. The brass
arc (radius over 6 ft.) could be read to 1/60 degree.
This sketch, from Tycho's own book, shows an observer
at F, a recorder, and a timekeeper with several clocks.
Good clocks had not been invented, but these were
the best Tycho could make.

who found the jump from Ptolemy to Copernicus
too big. He pictured the five planets (without the
Earth) all moving in circles around the Sun. Then
the whole group, Sun and planets, moved around
the Earth; and so did the Moon. Geometrically, this
is equivalent to the Copernican scheme, but it avoids
the uncomfortable feeling of a moving Earth.

Tycho became the foremost man of science in
Europe. Philosophers, statesmen, even kings, and
many scientists, came to visit him. They were received in grand style and shown the wonders of
the castle and its instruments. Yet Tycho could be
hot tempered and haughty to people he thought
stupid or visiting only for fashionable gossip. To

such he appeared to be a rude contradictory little
man with a violent temper, but to the wise he was
a great experimental scientist with a passion for
accuracy and a delight in marvels.

For all his scientific fervor, Tycho was vain and
superstitious. He kept a half-witted dwarf in his
household; and at banquets, with his peasant-born
wife presiding, Tycho insisted on listening to the
dwarf's remarks as prophetic. "It must have been
an odd dinner party, with this strange, wild, terribly
clever man, with his red hair and brazen nose, sometimes flashing with wit and knowledge, sometimes
making the whole company, princes and servants
alike, hold their peace and listen humbly to the rav-
ings of a poor imbecile."[6]

Troubles

While Tycho's grand observatory attracted visitors
from far and wide, his impetuous ways brought him
troubles. He made jealous enemies at court, and he
had serious troubles with his tenants. When the
King had given Tycho the island for life, the peas-
ants who had small farms on it were bound to do
some work for him as his tenants. They did much
of the work of building Uraniborg, and after that

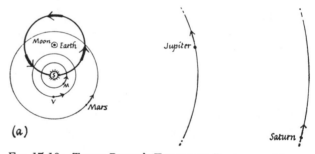

FIG. 17-10a. TYCHO BRAHE'S THEORY OF PLANETARY MOTION
The Sun moves around a fixed Earth and carries
all the rest of the Copernican system with it.

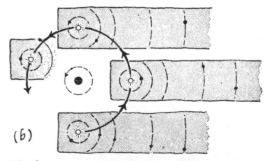

FIG. 17-10b. TYCHO BRAHE'S THEORY OF PLANETARY MOTION
A sketch, *not to scale*, showing successive positions
of the system in January, April, July, September.
(The planetary system moves like a frying-pan
given a circular motion by a housewife who wants
to melt a piece of butter in it quickly.)

[6] Professor Stuart quoted by Sir Oliver Lodge.

FIGS. 17-6, 7, 8, 9. SOME OF TYCHO'S INSTRUMENTS IN URANIBORG.
Reprinted from *Tycho Brahe's Description of His Instruments and Scientific Work* by permission of The Royal Danish Academy of Sciences and Letters.

ARMILLÆ ZODIACALES

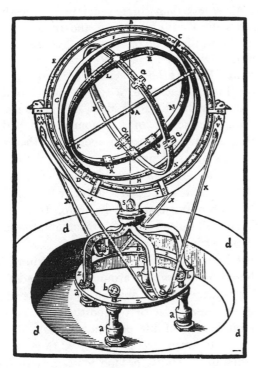

FIG. 17-6. AN "ASTROLABE" BUILT BY TYCHO, following the design used by Hipparchus. This instrument measures the latitude and longitude of a star or planet directly. (Diameter of circles: about 4 ft.) Tycho built several improved forms, with one axis parallel to the Earth's polar axis.

SEXTANS ASTRONOMICUS TRIGO-
NICUS PRO DISTANTIIS RIMANDIS

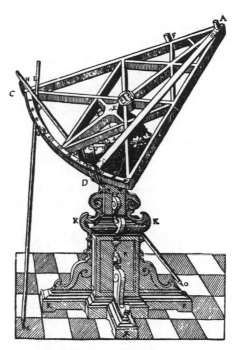

FIG. 17-7. ONE OF TYCHO'S SEXTANTS
This instrument, with brass scales and wooden frame, was used to measure the angle between the directions of two stars, by two observers sighting simultaneously along arms AD, AC. It was carried on a globe which could twist in firm supports, so that it could be tilted in any direction. (Length of arms: about 5 ft. Angles estimated to $\frac{1}{240}°$.)

QUADRANS MAGNUS CHALIBEUS.
IN QUADRATO ETIAM CHALIBEO COMPREHENSUS,
UNAQUE AZIMUTHALIS

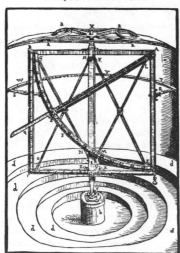

FIG. 17-8. TYCHO'S GREAT QUADRANT
Radius about 6 ft.

PARALLATICUM ALIUD SIVE REGULÆ
TAM ALTITUDINES QUAM AZIMUTHA
EXPEDIENTES

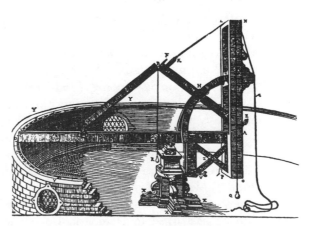

FIG. 17-9. A RULER BUILT BY TYCHO,
to measure altitudes and azimuths.

Tycho made them do chores for his household. He was unreasonable in his demands and haughty in his treatment—he may even have put unwilling workers in his prison. Several times complaints reached the King, who had to intervene. The King had given Tycho other estates as well, whose rents were to provide him with money for living and running the observatory. In return, Tycho was expected to keep the estates in reasonable repair; and, again, complaints reached the King that he failed to do so.

King Frederik's intense interest in Tycho's work protected him, but when the King died, eleven years after the observatory was built, Tycho's troubles began to grow. Young King Christian IV came to the throne and was surrounded by nobles who were less favorable to Tycho. Some of Tycho's estates were withdrawn, and he began to worry about the future. He wrote to a friend saying he might have to leave the island, comforting himself that "every land is home to a great man," and that wherever he went the same heavens would be over his head. The young King was sympathetic but needed to economize. More of Tycho's estates were taken. Tycho, seeing that he would not have enough money to maintain Uraniborg, moved to the mainland. Feeling unwelcome there, he decided to leave his ungrateful country and look for a new patron and place to work. He took his smaller instruments with him and set them up in temporary quarters in Germany while he negotiated with the Emperor Rudolph of Bohemia, an enlightened ruler with a great interest in science. He wrote a long haughty letter to young King Christian of Denmark offering to return, and received only a chilly reply. Meanwhile, he printed a large illustrated catalogue of his instruments and sent elegantly bound copies to possible patrons, including Rudolph.

The New Observatory in Prague

Finally, after two years of travelling and visiting, he arrived in Prague and was welcomed by Rudolph, who gave him a castle for observatory and promised him a huge salary. The Emperor was genuinely interested in astronomy (and probably astrology too), but he was careless as a ruler and could not always pay Tycho in full. Yet he did re-establish Tycho, and he deserves great credit for thus saving Tycho's work—credit which he received when Tycho's records were published with the name "Rudolphine Tables."

In his new castle, Tycho lived the same weird, earnest life he had lived in Denmark. He fetched his big instruments across Germany from Denmark and gathered round him a small school of astronomers and mathematicians. But his spirit was broken; he was a stranger in a strange land. He continued his observations and began setting up the Rudolphine Tables, but he became more and more despondent. After less than three years in Prague, he was seized with a painful disease and died. In the delirium of his illness he often cried, "Ne frustra vixisse videar," "Oh that I may not appear to have lived in vain." His life had been given him not just to enjoy but to achieve a great work and, still yearning for this, his life's ideal, he died. This doubt was undeserved by an astronomer who had catalogued a thousand stars so accurately that his observations are still used, a man who recorded the planets' positions for twenty years with accuracy calculated to $\frac{1}{60}$ of a degree, a man who gave Kepler and Newton the essential basis for their work in turn. He had succeeded in his original intention, and his work was not in vain.

Just before he died, free from delirium, Tycho gathered his household around him, asked them to preserve his work, and entrusted one of his students, Johannes Kepler, with the editing and correcting and publishing of his planetary tables. The great instruments were preserved for a time but were smashed in the course of later warfare: only Tycho's great celestial globe now remains. The island observatory was broken up and there is little sign of it today. Denmark lost its great name as a center of science; and it was not until this century that its fame grew again as such a world center, this time around the name of Niels Bohr.

PROBLEMS FOR CHAPTER 6

★ 1. TYCHO'S PRECISION

Tycho made some of his observations with plumb lines and peepholes like rifle sights. His final estimates were usually to be trusted to 1 minute of angle ($= \frac{1}{60}$ of 1 degree). To see how careful he must have been, answer the following questions:

(a) Suppose he had pointed his peephole sights at a planet, and that they carried a graduated angle-scale with them on which he could read the position of a vertical plumb line. Suppose his angle-scale was part of a circle of radius 7 feet. (Your protractor in an ordinary box of geometrical instruments has a radius about 3 inches.) How thin must the thread of his plumb line have been so that a mistake of 1 thread-thickness on the scale made an error of 1 minute of angle? (Give your answer as a fraction of an inch.) (*Hint*: With $r = 7$ ft, length of whole scale of 360° round the circumference would be . . . . Then 1° must take a length of scale about . . . . Then 1 minute must take . . . .)

(b) Does your estimate call for a cord, a string, a thread, or a spider filament?

2. Why did Kings often support Astronomers?

★ 3. PARALLAX AND STARS

(a) Suppose all the stars in some group or constellation were *infinitely* far away except one star and that single star was only a few billion miles away. What would its parallax motion look like? How long would the motion take for one cycle? Describe the pattern of this (apparent) motion for a close star (i) near the ecliptic; (ii) near the pole of the ecliptic (90° from ecliptic).

(b) To see whether Tycho Brahe had a hope of detecting the tiny parallax motion of the nearest fixed stars try the following calculation. In 6 months the Earth swings around half its orbit from one end of a diameter to the other, 186,000,000 miles, *straight across* from A to B. (Fig. 17-11) Suppose Tycho looked at a very near star, S, against a background

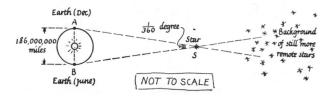

FIG. 17-11. PROBLEM 2.
(Diagram *not to scale*—the real angle is only ⅟₃₆₀°)

of other stars which are very much farther away. In watching the position of S against the background, Tycho would swing his sight-line through an angle ASB which he would measure as an angular displacement of star S, among the background stars. Suppose this angle to be ⅟₃₆₀ of a degree. It seems doubtful that he could have detected a smaller shift than that. Use rough arithmetic ("judging") to answer the following:

(i) Taking the angle ASB to be ⅟₃₆₀ of a degree, estimate the distance, AS, of the star from the Earth. Use a method like that of Problem 1 above—do not try to use trigonometry for these extreme angles. (Take the shallow arc, AB, with center at S, as 186,000,000 ≈ 200,000,000 miles.)

(ii) Compare the result of (i) above with modern measurements. These are usually expressed by giving the time light takes to travel the distance considered. Light travels the *diameter* of the Earth's orbit, AB, in about 16 minutes (8 minutes from the Sun to us). Light from the nearest star takes about 4 years (≈ 2,000,000 minutes) to reach us. What value does *this* give for the angle ASB? (Avoid trig. Argue simply by proportions.)

CHAPTER 7 · JOHANNES KEPLER (1571-1630)

~~~~~~

> Out of the night that covers me,
> Black as the pit from pole to pole
> I thank whatever gods there be
> For my unconquerable soul.

> —W. E. Henley (in hospital, 1875)

"For there is a musick where ever there is a harmony, order, or proportion: and thus far we may maintain the musick of the sphears; for those well-ordered motions and regular paces, though they give no sound unto the ear, yet to the understanding they strike a note most full of harmony."

> From Religio Medici
> Sir Thomas Browne (1642)

~~~~~~

Kepler, the young German to whom Tycho Brahe left his tables, was well worthy of this trust. He grew into one of the greatest scientists of the age—perhaps equalled in his own time only by Galileo and later outshone only by Newton. As Sir Oliver Lodge points out, Tycho and Kepler form a strange contrast: Tycho "rich, noble, vigorous, passionate, strong in mechanical ingenuity and experimental skill, but not above the average in theoretical power and mathematical skill"; and Kepler "poor, sickly, devoid of experimental gifts, and unfitted by nature for accurate observation, but strong almost beyond competition in speculative subtlety and innate mathematical perception."[1] Tycho's work was well supported by royalty, at one time magnificently endowed; Kepler's material life was largely one of poverty and misfortune. They had in common a profound interest in astronomy and a consuming determination in pursuing that interest.

Kepler was born in Germany, the eldest son of an army officer. He was a sickly child, delicate and subject to violent illnesses, and his life was often despaired of. The parents lost their income and were reduced to keeping a country tavern. Young Johannes was taken from school when he was nine and continued as a servant till he was twelve. Ultimately he returned to school and went on to the University where he graduated second in his class. Meanwhile, his father abandoned his home and returned to the army; and his mother quarreled with her relations, including her son, who was therefore glad to get away. At first he had no special interest in astronomy. At the University he heard the Coper-

nican system expounded. He adopted it, defended it in a college debate, and even wrote an essay on one aspect of it. Yet his major interests at that time seem to have been in philosophy and religion, and he did not think much of astronomy. But then an astronomical lecturership fell vacant and Kepler, who was looking for work, was offered it. He accepted reluctantly, protesting, he said, that he was not thereby abandoning his claim "to be provided for in some more brilliant profession." In those days astronomy had little of the dignity which Kepler himself later helped to give it. However, he set to work to master the science he was to teach; and soon his learning and thinking led to more thinking and enjoyment. "He was a born speculator just as Mozart was a born musician"[1]; and he *had* to find the mathematical scheme underlying the planetary system. He had a restless inquisitive mind and was fascinated by puzzles concerning numbers and size.[2] Like Pythagoras, he "was convinced that God created the world in accordance with the principle of per-

[2] Most of us have similar delights, though less intense. You have probably enjoyed working on series of numbers, given as a puzzle or an "intelligence test," trying to continue the series. Try to continue each of the following. If you enjoy puzzling over them (as well as succeeding) you are tasting something of Kepler's happiness.

(a) 1, 3, 5, 7, 9, 11, . . . How does this series probably go on?
(b) 1, 4, 9, 16, 25, . . . ?
(c) 5, 6, 7, 10, 11, 12, 15, 16, . . . ?
(d) 2, 3, 4, 6, 8, 12, 14, 18, . . . ?
(e) 4, 7, 12, 19, 28, . . . ?
(f) 1 7 3 6 5 5 7 4 9 . . . ?
(g) 0 1 8 8 1 1 0 2 4 1 5 6 2 5 . . . ?

[Note that in (f) and (g) you must also find where to put the commas.]

[1] Sir Oliver Lodge, *Pioneers of Science.*

Fig. 18-1. ? Law Relating Sizes of Planetary Orbits ?

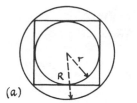
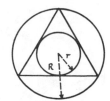

(a)

Fig. 18-1a. Kepler's First Guess
A regular plane figure (such as a square) can have a circle inscribed, to touch its sides. It can also have an outside circle, through its corners. Then that outside circle can be the inner circle for another, larger plane figure. The ratio of radii, R/r, is the same for all squares; and it has a different fixed value for all triangles. Geometrical puzzle: what is the fixed value of R/r for the inner and outer circles of a square? What is the value for a triangle?

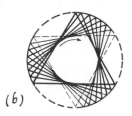

(b)

Fig. 18-1b. The same two circles can be generated by letting the figure (here a triangle) spin around its own center, in its own plane. Its corners will touch the outer circle, and its sides envelop the inner one.

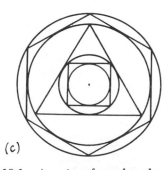

(c)

Fig. 18-1c. A series of regular plane figures, separated by inner and outer circles, provides a series of circles which might show the proportions of the planetary orbits. Even the best choice of figures failed to fit the solar system.

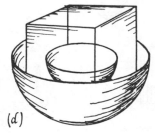

(d)

Fig. 18-1d. Kepler's Second Guess
This shows the basis of Kepler's final scheme. He chose the order of regular solids that gave the best agreement with the known proportions of planetary orbits.

fect numbers, so that the underlying mathematical harmony . . . is the real and discoverable cause of the planetary motions."[3] Kepler himself said, "I brooded with the whole energy of my mind on this subject."

His mind burned with questions: Why are there only *six* planets? Why do their orbits have just the proportions and sizes they do? Are the times of the planets' "years" related to their orbit-sizes? The first question, "Why just six?" is characteristic of Kepler's times—nowadays we should just hunt for a seventh. But then there was a finality in facts and a magic in numbers. The Ptolemaic system counted seven planets (including Sun and Moon, excluding the Earth) and even had arguments to prove seven must be right.

Kepler tried again and again to find some simple relation connecting the radius of one orbit with the next. Here are *rough relative* radii from Tycho's observations, calculated for the Copernican scheme: 8 : 15 : 20 : 30 : 115 : 195. He tried to guess the secret in these proportions. Each guess meant a good deal of work, and each time he found it did not fit the facts he rejected that guess honestly. His mystical mind clung to the Greek tradition that circles are perfect; and at one time he thought he could construct a model of the orbits thus: draw a circle, inscribe an equilateral triangle in it, inscribe a circle in that triangle, then another triangle inside the inner circle, and so on. This scheme gives successive circles a definite ratio of radii, 2:1. He hoped the circles would fit the proportions of the planetary orbits if he used squares, hexagons, etc., instead of some of the triangles. No such arrangement fitted. Suddenly he cried out, "What have flat patterns to do with orbits in *space*? Use solid figures." He knew there are only five completely regular crystalline solid shapes (see Fig. 18-3). Greek mathematicians had proved there cannot be more than five. If he used these five solids to make the separating spaces between six spherical bowls, the bowls would define six orbits. Here was a wonderful reason for the number six. So he started with a sphere for the Earth's orbit, fitted a dodecahedron outside it with its faces touching the sphere, and another sphere outside the dodecahedron passing through its corners to give the orbit of Mars; outside that sphere he put a tetrahedron, then a sphere for Jupiter, then a cube, then a sphere for Saturn. Inside the Earth's sphere he placed two more solids separated by spheres, to give the orbits of Venus and Mercury.

[3] Sir William Dampier, *History of Science* (4th edn., Cambridge University Press, 1949).

74

THE REGULAR SOLIDS. A geometrical intelligence test

How many different shapes of regular solid are possible? To find out, follow argument (a); then try (b).

A regular solid is a geometrical solid with identical regular plane faces; that is, a solid that has:

all its edges the same length
all its face angles the same
all its corners the same
and all its faces the same shape.

(See opposite for shapes that do not meet the requirements.)

FIG. 18-2.

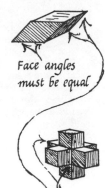

For example, a cube is a regular solid.

The faces of a regular solid might be:

all equilateral triangles
or all squares
or all regular pentagons
or . . . and so on . . .

&c.

(a) Here is the argument for square faces. Try to make a corner of a regular solid by having several corners of squares meeting there.

We already know that in a cube each corner has three square faces meeting there. Take three squares of cardboard and place them on the table like this, then try to pick up the place where three corners of squares meet. The squares will fold to make a cube corner.

Therefore we can make a regular solid with three square faces meeting at each of the solid's corners. (We need three more squares to make the rest of the faces and complete the cube.)

Could we make another regular solid, with only one, or two, or four square faces meeting at a corner?

With *one square*, we cannot make a solid corner.

With *two squares*, we can only make a flat sandwich.

With *three squares*, we make a cubical corner, leading to a cube.

With *four squares* meeting at a corner, they make a flat sheet there, and cannot fold to make a corner for a closed solid.

Thus, SQUARES CAN MAKE ONLY *ONE* KIND OF REGULAR SOLID, A CUBE.

(b) Now try for yourself with regular pentagons, and ask how many regular solids can be made with such faces.

Then try hexagons, and other polygons.

Then return to triangles and carry out similar arguments with triangular faces.

THE RESULT: Only FIVE varieties are possible in our 3-dimensional world. (Fig. 18-3)

(NOTE that these arguments need pencil sketches but can be carried out in your head without cardboard models.)

THE SOLIDS BELOW ARE NOT REGULAR SOLIDS

Edges must be equal

Face angles must be equal

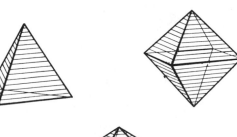

All faces must be same

All corners must be same

THE REGULAR SOLIDS

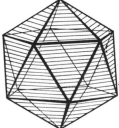

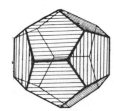

FIG. 18-3.
The five regular solids are drawn after D. Hilbert and S. Cohn-Vossen in *Anschauliche Geometrie* (Berlin: Julius Springer, 1932).

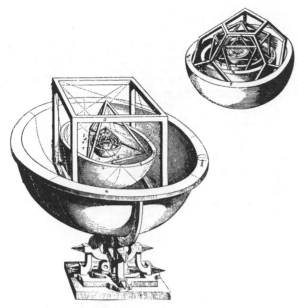

FIG. 18-4. KEPLER'S SCHEME OF REGULAR SOLIDS, FROM HIS BOOK

The relative sizes of planetary orbits were shown by bowls separating one solid from the next. The bowls were not thin shells but were just thick enough to accommodate the *eccentric* orbits of the planets.

The relative radii of the spheres, calculated by geometry, agreed fairly well with the proportions then known for planetary orbits, and Kepler was overjoyed. He said: "The intense pleasure I have received from this discovery can never be told in words. I regretted no more the time wasted; I tired of no labor; I shunned no toil of reckoning, days and nights spent in calculation, until I could see whether my hypothesis would agree with the orbits of Copernicus, or whether my joy was to vanish into air."

We now know the scheme was only a chance success. In later years, Kepler himself had to juggle the proportions by thickening up the bowls to fit the facts; and, when more planets were discovered centuries after, the scheme was completely broken.[4] Yet this "success" sent Kepler on to further, great discoveries.

He published his discovery in a book, including an account of all his unsuccessful trials as well as the successful one. This unusual characteristic appeared in many of his writings. He showed *how* his discoveries were made. He had no fear of damaging his reputation but only wanted to increase human knowledge, so instead of concealing his mistakes he gave a full account of them. "For it is my opinion,"

[4] There *is* a rough empirical rule relating orbit-radii to each other, called Bode's Law; but until recently no reason for it could be found. However, see G. Gamow, *1, 2, 3, . . . Infinity* (New York, Mentor Books, 1953) for a suggested reason.

he said, "that the occasions by which men have acquired a knowledge of celestial phenomena are not less admirable than the discoveries themselves. . . . If Christopher Columbus, if Magellan, if the Portuguese when they narrate their *wanderings*, are not only excused, but if we do not wish these passages omitted, and should lose much pleasure if they were, let no one blame me for doing the same."

The book also contained an admirable defense of the Copernican system, with good solid reasons in its favor. Young Kepler sent copies of his book to Tycho Brahe and Galileo, who praised it as a courageous beginning. This started Kepler's lifelong friendship with them.[5] In the same book, he made the suggestion that each planet may be pushed along in its orbit by a spoke carrying some influence from the Sun—a vague and improbable idea that later helped him discover his second Law.

Kepler was a Protestant, and he found himself being turned out of his job by Roman Catholic pressure on the administration. Worrying about his future, and anxious to consult Tycho on planetary observations, he travelled across Germany to Prague. Tycho, busy observing Mars, "the difficult planet," wrote to him: "Come not as a stranger but as a friend; come and share in my observations with such instruments as I have with me." While the work of the observatory proceeded, Tycho was turning to detailed "theory," schemes to fit his long series of observations. Kepler was soon set to work on Mars, working with Tycho to find a circular orbit that fitted the facts. Sensitive, and sick, Kepler complained that Tycho treated him as a student and did not share his records freely. Once, driven half crazy by worry, he wrote Tycho a violent letter full of quite unjust reproaches, but Tycho merely argued gently with him. Kepler, repenting, wrote:

"Most Noble Tycho,

How shall I enumerate or rightly estimate your benefits conferred on me? For two months you have liberally and gratuitously maintained me, and my whole family . . . you have done me every possible kindness; you have communicated to me everything you hold most dear. . . . I cannot reflect without consternation that I should have been so given up by God to my own intemperance as to shut my eyes

[5] In a later edition, Kepler took special trouble to avoid any appearance of stealing credit from Galileo. In one of his rejected theories he assumed a planet between Mars and Jupiter. Fearing a careless reader might take this to be a claim anticipating Galileo's discovery of Jupiter's moons, he added a note, saying of his extra planet, "Not circulating round Jupiter like the Medicaean stars. Be not deceived. I never had them in my thoughts."

on all these benefits; that, instead of modest and respectful gratitude, I should indulge for three weeks in continual moroseness towards all your family, in headlong passion and the utmost insolence towards yourself. . . . Whatever I have said or written . . . against your excellency . . . I . . . honestly declare and confess to be groundless, false, and incapable of proof."

When Kepler ended his visit and returned to Germany, Tycho again invited him to join him permanently. Kepler accepted but was delayed by poverty and sickness, and when he reached Prague with no money he was entirely dependent on Tycho. Tycho secured him the position of Imperial Mathematician to assist in the work on the planets.

Tycho died soon after, leaving Kepler to publish the tables. Though he still held the imperial appointment, Kepler had difficulty getting his salary paid and he remained poor, often very poor. At one time he resorted to publishing a prophesying almanac. The idea was abhorrent to him, but he needed the money, and he knew that astrology was the form of astronomy that would pay. For the rest of his life, over a quarter of a century, he worked on the planetary motions, determined to extract the simple secrets he was sure must be there.

The Great Investigation of Mars

When Tycho died, Kepler had already embarked on his planetary investigations, chiefly studying the motion of Mars. What scheme would predict Mars' orbit? Still thinking in terms of circles, Kepler made the planet's orbit a circle round the Sun, with the Sun a short distance off center (like Ptolemy's eccentric Earth). Then he placed an equant point Q off center on the other side, with a spoke from Q to swing the planet around at constant speed. He did not insist, like Ptolemy, on making the eccentric distances CS and CQ equal, but calculated the best proportions for them from some of Tycho's observations. Then he could imagine the planet moving around such an orbit and compare other predicted positions with Tycho's record. He did not know the direction of the line SCQ in space, so he had to make a guess and then try to place a circular orbit on it to fit the facts. Each trial involved long tedious calculations, and Kepler went through 70 such trials before he found a direction and proportions that fitted a dozen observed longitudes of Mars closely. He rejoiced at the results, but then to his dismay the scheme failed badly with Mars' latitudes. He shifted his eccentric distances to a compromise value to fit the latitudes; but, in some parts of the orbit,

Mars' position as calculated from his theory disagreed with observation by 8' (8 sixtieths of one degree). Might not the observations be wrong by this small amount? Would not "experimental error" take the blame? No. Kepler knew Tycho, and he was sure Tycho was never wrong by this amount. Tycho was dead, but Kepler trusted his record. This was a great tribute to his friend and a just one. Faithful to Tycho's memory, and knowing Tycho's methods, Kepler set his belief in Tycho against his own hopeful theory. He bravely set to work to go the whole weary way again, saying that upon these eight minutes he would yet build up a theory of the universe.

It was now clear that a circular orbit would not do. Yet to recognize any other shape of orbit he must obtain an accurate picture of Mars' real orbit from the observations—not so easy, since we only observe the apparent path of Mars from a moving Earth. The true distances were unknown; only angles were measured and those gave a foreshortened compound of Mars' orbital motion and the Earth's. So Kepler attacked the Earth's orbit first, by a method that had all the marks of genius.

Mapping the Earth's Orbit in Space and Time

To map the Earth's orbit around the Sun on a scale diagram, we need many sets of measurements, each set giving the Earth's bearings from *two* fixed points. Kepler took the fixed Sun for one of these, and for the other he took Mars *at a series of times when it was in the same position in its orbit.* He proceeded thus: he marked the "position" of Mars in the star pattern at one opposition (opposite the Sun, overhead at midnight). That gave him the direction of a base line Sun-(Earth)-Mars, SE_1M. Then he turned the pages of Tycho's records to a time *exactly one Martian year later.* (That time of Mars' motion around its orbit was known accurately, from records over centuries.) Then he knew that Mars was in the same position, M, so that SM had the same direction. By now, the Earth had moved on to E_2 in its orbit. Tycho's record of the position of the Mars in the star-pattern gave him the new apparent direction of Mars, E_2M; and the Sun's position gave him the direction E_2S. Then he could calculate the angles of the triangle SE_2M from the record, thus: since he knew the directions E_1M and E_2M (marked on the celestial sphere of stars) he could calculate the angle A between them. Since he knew the directions E_1S and E_2S, he could calculate the angle B, between them. Then on a scale diagram he could choose two points to represent S and M and

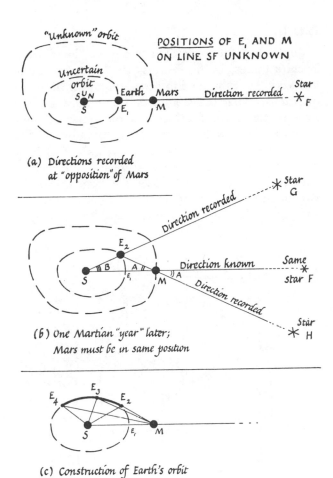

(a) Directions recorded
at "opposition" of Mars

(b) One Martian "year" later;
Mars must be in same position

(c) Construction of Earth's orbit

FIG. 18-5. KEPLER'S SCHEME TO PLOT THE EARTH'S ORBIT

locate the Earth's position, E_2, as follows: at the ends of the fixed baseline SM, draw lines making angles A and B and mark their intersection E_2. One Martian year later still, he could find the directions E_3M and E_3S from the records, and mark E_3 on his diagram. Thus Kepler could start with the points S and M and locate E_2, E_3, E_4, . . . enough points to show the orbit's shape.*

Then, knowing the Earth's true orbit, he could invert the investigation and plot the shape of Mars' orbit. He found he could treat the Earth's orbit either as an eccentric circle or as slightly oval; but Mars' orbit was far from circular: it was definitely oval or, as he thought, egg-shaped, but he still could not find its mathematical form.

Variable Speed of Planets: Law II

Meanwhile his plot of the Earth's motion in space showed him just how the Earth moves unevenly along its orbit, faster in our winter than in summer.

---

* Einstein spoke of this method as one of Kepler's greatest achievements.

He sought for a law of uneven speed, to replace the use of the equant. His early picture of some pushing influence from the Sun suggested a law to try. He believed that motion needed a force to maintain it, so he pictured a "spoke" from the Sun pushing each planet *along* its orbit, a weaker push at greater distance. He tried (with a confused geometrical scheme) to add up the effects of such pushes from an eccentric Sun; and he discovered a simple law: *the spoke from Sun to planet sweeps out equal areas in equal times.* It does not swing around the Sun with constant speed (as Ptolemy would have liked), but it does have a constancy in its motion: constant rate of sweeping out area (which Ptolemy would probably have accepted). Look at the areas for equal periods, say a month each. When the planet is far from the Sun the spoke sweeps out a long thin triangle in a month; and as the planet approaches the Sun the triangles grow shorter and fatter—the planet moves faster. Later on, when Kepler knew the shape of Mars' orbit he tried the same rule and found it true for Mars too. Here he had a simple law for planetary speeds: each planet moves around the Sun with such speeds that the radius from Sun to planet sweeps out equal areas in equal times. Kepler had only a vague "reason" for it, in terms of solar influences, perhaps magnetic; but he treasured it as a true, simple statement, and used it in later investigations. We treasure it too, and assign a first-class reason to it. We call it Kepler's Second Law. His First Law, discovered soon after, gave the true shape of planetary orbits.

The Orbit of Mars: Law I

When he had plotted Mars' orbit (forty laboriously computed points), Kepler tried to describe its oval shape mathematically. He had endless difficulties—at one time he says he was driven nearly out of his mind by the frustrating complexity. He wrote to the Emperor (to encourage finances), in

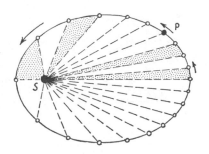

FIG. 18-6. KEPLER'S DISCOVERIES FOR MARS

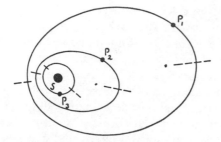

FIG. 18-7. A SOLAR SYSTEM WITH ELLIPTICAL ORBITS
AROUND A COMMON SUN
(The planets' orbits in our own Solar System have much
smaller eccentricities. But some comets move in
elliptical orbits with great eccentricity.)

his grandiose style: "While triumphing over Mars, and preparing for him, as for one already vanquished, tabular prisons and equated excentric fetters, it is buzzed here and there that my victory is vain, and that the war is raging anew. For the enemy left at home a despised captive has burst all the chains of the equations, and broken forth from the prisons of the tables."

Finally, he found the true orbit sandwiched between an eccentric circle that was too wide and an inscribed ellipse that was too narrow. Both disagreed with observation, the circle by +8' at some places, the inner ellipse by —8'. He suddenly saw how to compromise half way between the two, and found that gave him an orbit that is *an ellipse with the Sun in one focus.* He was so delighted with his final proof that this would work that he decorated his diagram with a sketch of victorious Astronomy (Fig. 18-8). At last he knew the true orbit of Mars.[6] A similar rule holds for the Earth and other planets. This is his First Law.

[6] It may seem strange that he did not think of an ellipse earlier. It was a well-known oval, studied by the Greeks as one of the sections of a cone. But then *we* know the answer. Besides, ellipses were not so important then. It was Kepler who added greatly to their fame. (An ellipse is easy to draw with a loop of string and two thumb-tacks. If you have never tried making one for yourself you should do so. This is an amusing experiment which will show you a property of ellipses that is valuable in optics.)

FIG. 18-9.
DRAWING AN ELLIPSE, with a loop of thread and two nails

Law III

Kepler had then extracted two great "laws" from Tycho's tables, by his fearless thinking and untiring work. He continued to brood on one of his early questions: what connection is there between the sizes of the planets' orbits and the times of their "years"? He now knew the average radii[7] of the orbits; the times of revolution ("years") had long been known. (As the Greeks surmised, the planets with the longest "years" have the largest orbits.) He felt sure there was some relation between radius and time. He must have made and tried many a

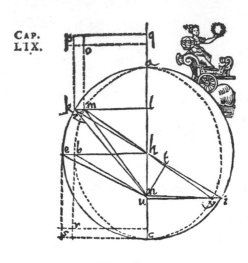

FIG. 18-8.
KEPLER'S TRIUMPHANT DIAGRAM, FROM HIS BOOK ON MARS
When he succeeded in proving that an ellipse with the
Sun in one focus could replace an oscillating circular
orbit and maintain an "equal area" law, Kepler added
a sketch of Victorious Astronomy, to show his delight
and to emphasize the importance of the proof.

guess, some of them sterile ones like his early scheme of the five regular solids or wild mystical ones like his speculation of musical chords for the planets. Fortunately there *is* a connection between radii and times, and Kepler lived to experience the joy of finding it. He found that the fraction R^3/T^2 is the same for all the planets, where R is the planet's average orbit-radius, and T is the planet's "year," measured in our days. See the table.

[7] Assuming circular orbits, Copernicus made rough estimates, and Tycho made better ones. Kepler knew these when he tried his strange scheme of regular solids, and he traded on their roughness to let his test of that theory seem "successful."

PLANETARY DATA — TEST OF KEPLER'S THIRD LAW
(These are modern data, more accurate than Kepler's)

Planet	Radius of planet's orbit R (miles)	Time of revolution (planet's "year") T (days)	R^3 (miles)^3	T^2 (days)^2	$\dfrac{R^3}{T^2}$ $\dfrac{(miles)^3}{(days)^2}$
Mercury	5.785×10^7	87.97	1.936×10^{23}	7739.	2.502×10^{19}
Venus	10.81×10^7	224.7	12.63×10^{23}	50490.	2.501×10^{19}
Earth	14.95×10^7	365.3	33.41×10^{23}	133400.	2.504×10^{19}
Mars	22.78×10^7	687.1	118.21×10^{23}	472100.	2.504×10^{19}
Jupiter	77.76×10^7	4333.	470.18×10^{23}	18770000.	2.505×10^{19}
Saturn	142.58×10^7	10760.	2898.5×10^{23}	115800000.	2.503×10^{19}

The test of Kepler's guess is shown in the last column.

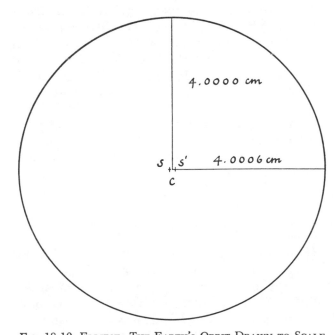

FIG. 18-10. ELLIPSE: THE EARTH'S ORBIT DRAWN TO SCALE

The actual eccentricity of planetary orbits is very small. The orbits are almost circles, yet Tycho's observations enabled Kepler to show that they are not circles but ellipses. The sketch above shows the Earth's orbit drawn to scale. If a 4.0000 centimeter line is used, as here, to represent the minimum radius, which is really some 93,000,000 miles, the maximum radius needs a line 4.0006 centimeters long. The eccentricity of Mars' orbit is over thirty times as big, but even then the ratio of radii is only 1.0043 to 1.0000. Mercury is the only planet with a much greater eccentricity of orbit, with radii in proportion 1.022 to 1.000. Even this eccentricity of orbit seems small, but it is sufficient to involve Mercury in such speed changes around the orbit that Relativity mechanics predicts a very slow slewing around of the orbit—a precession of only 1/80 of a degree per *century*, discovered and measured long before the Relativity prediction!

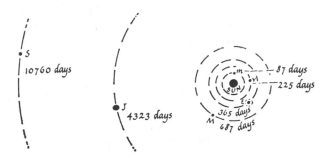

FIG. 18-11a. ? RELATIONSHIP BETWEEN *RADIUS* AND "*YEAR*" FOR PLANETARY ORBITS ?
(Planetary orbits roughly to scale.)

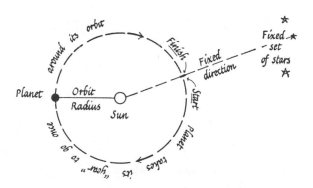

FIG. 18-11b. PLANET'S "YEAR"

The planet's year is the time it takes to go once around its orbit. This is the time-interval from the moment when its direction hits some standard mark in the star-pattern until it returns to the same mark. (The Earth moves too. An allowance for the Earth's motion must be made when extracting the planet's true year from observations.)

Again he was overjoyed at wresting a divine secret from Nature by brilliant guessing and patient trial. He said:

"What I prophesied two-and-twenty years ago, as soon as I discovered the five solids among the heavenly orbits—what I firmly believed long before I had seen Ptolemy's "Harmonies"—what I had promised my friends in the title of this book, which I named before I was sure of my discovery—what sixteen years ago, I urged as a thing to be sought— that for which I joined Tycho Brahe, for which I settled in Prague, for which I have devoted the best part of my life to astronomical contemplations, at length I have brought to light, and recognized its truth beyond my most sanguine expectations. It is not eighteen months since I got the first glimpse of light, three months since the dawn, very few days since the unveiled sun, most admirable to gaze upon, burst upon me. Nothing holds me . . . the die is cast, the book is written, to be read either now or by posterity, I care not which; it may well wait a century for a reader, as God has waited six thousand years for an observer."

Kepler's Laws

These investigations took years of calculating, changing, speculating, calculating. . . . Kepler discovered—among other "harmonies" that he valued— three great laws that are clear and true. Here they are:

LAW I   EACH PLANET MOVES IN AN ELLIPSE WITH THE SUN IN ONE FOCUS.

LAW II   THE RADIUS VECTOR (LINE JOINING SUN TO PLANET) SWEEPS OUT EQUAL AREAS IN EQUAL TIMES.

LAW III   THE SQUARES OF THE TIMES OF REVOLUTION (OR YEARS) OF THE PLANETS ARE PROPORTIONAL TO THE CUBES OF THEIR AVERAGE DISTANCES FROM THE SUN.
(Or R^3/T^2 is the same for all the planets)

Once guessed, the first two laws could be tested with precision with available data; so Kepler could make sure he had guessed right. Law III was tested in its discovery. Only relative values of orbit-radii were needed.

Kepler had done a great piece of work. He had discovered the laws that Newton linked with universal gravitation. Of course that was not what Kepler thought he was doing. "He was not tediously searching for empirical rules to be rationalised by a coming Newton. He was searching for ultimate causes, the mathematical harmonies in the mind of the Creator."[8] He emerged with no general reason for his ellipses and mathematical relationships; but he delighted in their truth.

Guessing the Right Law

Guessing the third law was a matter of finding a numerical relationship which would hold for several pairs of numbers. An infinite variety of "wrong" guesses can be made to fit a limited supply of data, in this case values of T and R for only six planets. Many such guesses that succeed with six planets fail when applied to a seventh planet (Uranus, discovered later). Of those that still succeed, many would fail if tried on an eighth planet (Neptune). So trials with more and more sets of data can help to remove "wrong" guesses, leaving the "right" one. But in what sense is the "right" one right? Some of us believe there is a really true story behind the things we see in Nature. Kepler, Galileo, and Newton probably thought like that. Others now say that the right rule is merely (a) *the rule that applies most generally* (for example, to the greatest variety of planets). In this sense Kepler's R^3/T^2 guess was right because it applies to later-discovered planets and to other systems such as Jupiter's moons. His five-regular-solids rule was wrong, because it did not agree well with data for the original six planets and failed completely when required to deal with more than six. And, they say, the right rule is (b) *the rule that fits best into a theoretical framework which ties together a variety of knowledge of Nature.* If that theory has been manufactured just to deal with the problem in hand, then (b) is nonsense—it would merely say that the rule is right because it agrees with its own theory constructed to agree with it. We call that an *ad hoc* theory. If, however, the theory connects the problem in hand to other natural knowledge, then (b) is a cogent recommendation. Newton, guessing at universal gravitation, made a theory that connects falling bodies and the Moon's motion and planetary motion and tides, etc. He showed that Kepler's Law III (as well as the other two) was a necessary deduction from this theory. Thus Kepler's R^3/T^2 rule seems "right" on both scores, (a) and (b), *generalness* and *agreement with wide theory.* It might have been a "wrong" guess, waiting like the early "five-regular-solids" law for more data to refute it and for theory to fail to "predict"[9] it.

[8] Sir William Dampier, *op.cit.*
[9] Scientists use "predict" in this way, but it is an unfortunate choice of word. Here it means "coordinate with other knowledge."

A Fictitious "Kepler Problem"

To see something of the hazards involved in an investigation like Kepler's let us trace through a specimen problem using imaginary data, with a fictitious relationship. Suppose you have invented a planetary puzzle and know the scheme you have used, but ask me to try to find the scheme. You present me with the following data.

	Data			Problem
"Planet"	R	T		What is the "law"
A	1	3		connecting R and T?
B	2	6		
C	4	18		

You know the scheme, since you have invented it. (It is not an inverse square law system: the "planets" are not real ones!) In fact, you got T by squaring R and adding 2. That is, you *chose* the relation $T = R^2 + 2$ and used it. (Make sure our data fit this formula.) So if a new planet D is discovered with $R = 5$ it will have $T = 5^2 + 2$, or 27. Suppose you give me the data for A, B, C (holding D up your sleeve). In looking for a rule, I try to find some algebraic combination of T and R which will be the *same for each of these planets*. Starting with planets A and B, I notice that T/R is 3/1 for A, 6/2 for B, the same for both. Hoping I have found the right rule (T/R the same for all), I try this on planet C. For C, T/R is 18/4 and this is not the same as 3/1. I must therefore reject this simple guess. In trying other schemes which give the same answer for planets A and B, I find several more which fail for C. But presently I find that I get the same answer for planets A and B if I proceed thus: I divide R into 8 and add 7 times R and subtract T; that is, I find the value of $8/R + 7R - T$.

For planet A, $8/1 + 7 \times 1 - 3 = 12$;
and for planet B, $8/2 + 7 \times 2 - 6 = 12$.

So the answer is the same, 12, for both A and B. Trying the same rule on planet C,

I have $8/4 + 7 \times 4 - 18 = 12$ again.

So I am delighted to find the rule works for C and A and B. Confident that I have got the right rule, I plan to publish it, but you then divulge the data for planet D: $R = 5$ and $T = 27$. Trying my rule on planet D,

I obtain $8/5 + 7 \times 5 - 27 = 9.6$.

After asking you whether your data might be wrong enough to excuse the difference between 9.6 and 12.0, I start all over again. If I am lucky as well as patient, I may hit upon a scheme such as this: add 2 to the square of R and divide by T. This yields an answer 1.000 for all four planets, A, B, C, D.[10] Therefore it has a better chance of being the right rule than the others. Tests on more data would improve its reputation further and if some general theory could endorse it I might feel sure I had the right rule. Summing up this investigation in a table, we have

---

[10] There is no special virtue in the answer being 1.000. If I divide by $5T$ instead of by T the answers would all be 0.200, but the essential story is unchanged.

"PLANET"	DATA		ATTEMPTS TO OBTAIN CONSTANT NUMBERS		
			1st Trial	Nth Trial	Qth Trial
	R	T	$\dfrac{T}{R}$	$\dfrac{8}{R} + 7R - T$	$\dfrac{R^2 + 2}{T}$
A	1	3	3	12	1
B	2	6	3	12	1
C	4	18	4.5	12	1
D	5	27	5.4	9.6	1
e	3	11	3.667	12.67	1

Note that at the last moment another "planet" has been discovered, e, which is so small that it was not noticed before. It too fits with the final rule (of course it does, in this game, since *you* manufactured its data by using your private knowledge of that rule), and it fails to fit with the earlier rules. Notice, however, that it nearly fits with the second rule, giving 12.67 instead of 12.00. If the data for planet e had been available when I was working on my second rule, should I not have been tempted to say "12.67 is near enough; the difference is due to experimental error"?

Kepler's Writing

Kepler wrote many books and letters setting forth his discoveries in detail, describing failures as well as successes. His account of his Laws is immersed in much mystical writing about other discoveries and ideas: planetary harmonies, schemes of magnetic influence, hints about gravitation, and a continuing delight in his earliest scheme of the five regular solids. Remember Kepler did not know the "right answers." He had no idea which of his theories would be validated by later discoveries and thought. He finally managed to get the Rudolphine tables printed—paying some of the cost himself, which he could hardly afford—so that at last really good astronomical data were available. Among his own books, he wrote a careful fairly popular book on general astronomy in which he explained the Copernican theory and described his own discoveries. The book was at once suppressed by the Church authorities, leaving him all the poorer by making it hard to get any of his books published and sold.

Comments on Kepler

"When Kepler directed his mind to the discovery of a general principle, he . . . never once lost sight of the explicit object of his search. His imagination, now unreined, indulged itself in the creation and invention of various hypotheses. The most plausible, or, perhaps, the most fascinating of these, was then submitted to a rigorous scrutiny; and the moment it was found to be incompatible with the results of observation and experiment, it was willingly abandoned, and another hypothesis submitted to the same severe ordeal. . . . By pursuing this method he succeeded in his most difficult researches, and discovered those beautiful and profound laws which have been the admiration of succeeding ages."[11]

Sir Arthur Eddington says:[12]

"I think it is not too fanciful to regard Kepler as in a particular degree the forerunner of the modern theoretical physicist, who is now trying to reduce the atom to order as Kepler reduced the solar system to order. It is not merely similarity of subject matter but a similarity of outlook. We are apt to forget that in the discovery of the laws of the solar system, as well as of the laws of the atom, an essential step was the emancipation from mechanical models. Kepler did not proceed by thinking out possible devices by which the planets might be moved across the sky— the wheels upon wheels of Ptolemy, or the whirling vortices of later speculation. I think that is how most of us would have attacked the problem; we should have hunted for some concrete mechanism to yield the observed motion, and have approached the laws of motion through an explanation of the motion. But Kepler was guided by a sense of mathematical form, an aesthetic instinct for the fitness of things. In these later days it seems to us less incongruous that a planet should be guided by the condition of keeping the Action stationary than that it should be pulled and pushed by concrete agencies. In like manner Kepler was attracted by the thought of a planet moving so as to keep the growth of area steady—a suggestion which more orthodox minds would have rejected as too fanciful. I wonder how this abandonment of mechanical conceptions struck his contemporaries. Were there some who frowned on these rash adventures of scientific thought, and felt unable to accept the new kind of law without any explanation or model to show how it could possibly be worked? After Kepler came Newton, and gradually mechanism came into predominance again. It is only in the latest years that we have gone back to something like Kepler's outlook, so that the music of the spheres is no longer drowned by the roar of machinery."

[11] Sir David Brewster, *Martyrs of Science*, 1848.
[12] In his Introduction to the Tercentenary Commemoration book on Kepler's life and work, *Johann Kepler*, Williams and Wilkins Company, for The History of Science Society, 1931.

Kepler carried astronomy through a great stage of development. His laws are landmarks in knowledge, rules true today for planetary systems and perhaps even for atomic models.

PROBLEM FOR CHAPTER 7

PUZZLE FOR A MODERN KEPLER

Radioactive atoms shoot out small atomic "projectiles," which are pieces of their own innermost core or nucleus. Many such atoms, including radium itself, shoot out a projectile that is itself an electrically charged helium atom (= a helium nucleus, = a helium atom stripped of its two electrons). These are called alpha-particles, or α-particles. The atomic "explosion" in which a radioactive atom shoots out an alpha-particle occurs spontaneously, the parent atom then changing into an entirely different kind of atom with different chemical properties. This is the characteristic of radioactivity. Such radioactive changes give us information about atomic structure. But they also provide "projectiles" which can be used to investigate the structure of other atoms, somewhat in the way in which a boxer investigates the structure of other boxers' faces. In particular a stream of alpha-particles was used to investigate the structure of atoms of gold in a very famous experiment which led to a revolutionary change in atomic theory. The problem below refers to that experiment. This and other tests of grand theoretical prediction are described in Chapter 40 of *Physics for the Inquiring Mind*.

A stream of α-particles was shot at a very thin leaf of gold in a vacuum. Most of them passed straight through, missing severe collisions with any gold atoms in the very thin leaf. But a few of the α-particles bounced out in new directions, having suffered severe collisions. A very few even bounced back. These observations suggested a new theory which then predicted just how many should bounce back in some chosen direction, out of every million fired. The theory predicted a definite relationship between the number of α-particles bouncing back (per million) and the speed with which they were travelling when they hit the gold leaf. The theory was tested by a crucial experiment, reported by Geiger and Marsden in *Philosophical Magazine*, Vol. 25, page 620, 1913. Some of the measurements are given below:

v Velocity of helium atoms (In "arbitrary units"*)	N Number of helium atoms bouncing back per minute in a standard chosen direction
2.00	25
1.91	29
1.70	44
1.53	81
1.39	101
1.13	255

* These velocities are in arbitrary units. One such unit was probably worth about 10,000,000 meters/second.

These data provide a problem somewhat like the one that faced Kepler when he had planetary orbit data but had not guessed his third law. There is a fairly simple relationship between N and v.

Can you find this relationship? Try this, as Kepler would, with courage and care, without any help from a theory or a book. If you find the relationship, show how closely the data fit it. Of course, the original experimenters had an advantage over you; they knew what relation to try first—but then they had to do a difficult experiment. In these difficult experiments of counting *single atoms* as they bounce away from the gold, you must not expect great accuracy; so, unlike Kepler's, your constant may wobble by 10% but not in any particular direction.

CHAPTER 8 · GALILEO GALILEI (1564-1642)

~~~~~~

"Science came down from Heaven to Earth on the inclined plane of Galileo."

~~~~~~

Galileo's life overlapped that of Kepler. When Tycho Brahe had moved to Prague with those instruments he had saved, and Kepler was starting his attack on Mars, Galileo, in his thirties, was growing famous as a mathematician and natural philosopher. In his life, Galileo did many great things for science; perhaps the greatest was establishing mathematical argument, tied to experiment, as the basis of scientific knowledge. He experimented, and he drew on the experiments of others, until he had an instinct for sound science; but he was above all a thinker and a teacher, and so good an arguer that he could out-argue the traditional philosophers on their own ground. He liked to use what we call his "thought experiments": hypothetical experiments devised for use in argument.[1] In these he appealed to common-sense knowledge of nature, or sometimes to specific experiments, and then argued out predictions of behavior or relationships. Thus, rather than call him the father of experimental science—as used to be the fashion—we might look on him as the first modern theoretical physicist.

Galileo gathered and taught the facts and ideas from which Newton formed his Laws of Motion. He drew on many contributions from earlier experimenters and thinkers—we even know, from the phrasing, which edition of certain earlier writers he copied in his books. He did not pull the new mechanics out of his hat, all his own discovery; but he did begin to build it into a comprehensive picture and he did make it *public* and *convincing*. He constructed one of the earliest telescopes and with it gathered new evidence to support Copernicus' theory and even Kepler's Third Law. He expounded the Copernican Theory with such compelling clearness that he upset traditional authorities. And he preached honest experiment and clear thinking with such exasperating fervour that he started physics on a new life.

Galileo and the New Science

Galileo's greatest contribution to the new physical science was a change of treatment. He brought back the scientific attitude of Pythagoras and Archimedes:

experimental knowledge should be codified by abstract mathematical ideas. For example, he stated clearly that for an object falling freely the distances fallen in times 1, 2, 3, 4, from rest run in the proportions 1:4:9:16: . . . (which we now express compactly by algebra: $s \propto t^2$). In stating this he cleared away modifications made by air-resistance, spinning, horizontal motion, etc., and described an ideal case for a particle falling in a vacuum. He derived, by simple mathematical reasoning, an alternative form: the distances fallen in successive equal intervals of time increase steadily, 1:2:3:4: . . . , or, as we now say, Δs (for $\Delta t = 1$) $\propto t$ (see note 2). Galileo, and his successors down to the present day, do not spoil science when they "think away" real conditions such as air-resistance. The modern scientist can formulate ideal mechanical laws for frictionless materials, weightless carts, unstretchable strings, . . . and then add the real conditions to modify the ideal laws.

Galileo also promoted a complete change of thought in astronomy: he broke the sharp distinction between heavenly affairs and earthly science. Copernicus had maintained the mystical ideal of perfect spheres, but Galileo tried to treat the planets and Sun and Moon as ordinary earthy bodies. He started applying the same treatment to a ball rolling downhill and a planet in the sky. He did not carry this democratic treatment through—he still endowed planets with a natural circular motion—yet he drew man's understanding of the whole universe towards a scheme of general mathematical laws.

For his mathematical treatment, Galileo had to deal with things that could be measured definitely. So he gave importance to "primary" qualities of matter, such as length, volume, velocity, force; and he disowned, as outside proper science, such subjective things as color, taste, smell, and musical sound, which he said, just disappear when the observer is not present.[3] Shakespeare hinted at this

[1] Example: His argument about three bricks falling.

[2] This is Galileo's simpler rule, for free fall with constant acceleration. If $\Delta t = 1$, Δs is the velocity; and this progression of Δs-values is a statement of constant acceleration—velocity increasing steadily with time.

[3] Such exclusions could be revoked if a scheme of measurement appeared. For example, the invention of a thermometer with a definite scale may bring our sense of warmth into good science.

(in *The Merchant of Venice*):

> The crow doth sing as sweetly as the lark,
> When neither is attended, . . .

Thus Galileo moved science towards the hardheaded mathematical treatment that followed Newton; and he carried philosophy towards the complete separation of matter and mind that followed Descartes. His teaching helped to make matter and motion seem true and real, while taste and color, etc., seem unreal, mere sensations in the observer's mind produced by shapes or motions of atoms—though those atoms themselves are well disciplined by mathematical laws. A century later Berkeley suggested that even the primary qualities of matter are unreal; they too come to our minds only as "sense-perceptions." On that view the whole organization of scientific laws and knowledge which Galileo helped to build is a framework of abstractions, a picture that we extract from the sense impressions the world sends us. It is a good picture, comforting, useful, interesting: but it is not the world itself. The world itself—whatever real or concrete world there is outside our senses—may well be far more complex than we can "know" in our scientific way. If we believe our scientific picture is completely real and true, we may find the laws of mechanics offering to trace the course of every atom from now into the future, and thereby threatening to predict all events, including our own decisions and actions. That would take away all our choice of action, all free will—a very distressing prospect. But that offer applies only to the abstract world of Newtonian science, not to the complex concrete world beyond. We should not let ourselves be frightened by the "fallacy of the misplaced concrete."[4]

Galileo's Life and Work—Pisa

Galileo was the son of an Italian nobleman who was himself a philosopher and musician. His home was in Pisa, near Florence. Though young Galileo wanted to be a painter, his father sent him to the University to study medicine, a field that was much respected and well paid. There, at the University of Pisa, he seized on a chance to learn geometry. (There is a story that he overheard a lecture on Euclid, was thrilled with it, and implored the lecturer to teach him.) His father opposed this new interest—mathematicians were poorly paid—but Galileo's enthusiasm could not be stopped. He devoured the works of Euclid, then read Archimedes, and soon started his own investigations of the properties of centers of gravity.

[4] A. N. Whitehead's phrase.

When he was 25, Galileo was appointed by the Duke, one of the ruling family of Medici, to the post of lecturer in mathematics—at a miserable salary. With great energy and zeal, but little tact, he set to work on the mechanics of moving bodies: reading the earlier books, sorting sense from nonsense, and putting statements and ideas to the test of experiment. He enjoyed annoying the Aristotelian philosophers around him by showing up the mistakes in their teaching. Though he was right and they were wrong, his tactless manner was not wise.

"The detection of long-established errors is apt to inspire the young philosopher with an exultation which reason condemns. The feeling of triumph is apt to clothe itself in the language of asperity, and the abettor of erroneous opinions is treated as a species of enemy to science. Like the soldier who fleshes his first spear in battle, the philosopher is apt to leave the stain of cruelty on his early achievements. . . . Galileo seems to have waged this stern warfare against the followers of Aristotle; and such was the exasperation which was excited by his reiterated and successful attacks, that he was assailed, during the rest of his life, with a degree of rancour which seldom originates in a mere difference of opinion."[5]

Galileo's realistic discussions of falling bodies and accelerated motion upset traditional teaching and were not welcome; nor were his arguments exposing the fallacies of old doctrines. While he gathered enthusiastic followers he also made enemies. Malice and jealousy made his position at Pisa so uncomfortable that he accepted an invitation to move to the University of Padua in the neighboring republic of Venice. At Padua, he found philosophers already talking of free fall as due to a force, and doubting whether it was wise science to rely on "natural places" or to look for "first causes." The time was ripe for Galileo's teaching. He taught with vigor and amazing skill, and he wrote on motion, mechanics, astronomy. Even then he was poorly paid. He had to run a lodging house for his students, and he set up an instrument-making shop.[6]

Padua

In his new post at Padua, his reputation grew. He loved to expound and argue. He was formidable in argument because he started by expounding his

[5] Sir David Brewster, *Martyrs of Science* (1848).

[6] He manufactured a "military compass" that combined the uses of a protractor and a slide rule, and received orders for it from many parts of Europe. He also sent some as presents to important people, to show how he could aid "the military art."

opponents' case more clearly than they could and then he demolished it—he was an intellectual truck-driver. He stayed at Padua twenty years, during which he gathered much knowledge of mechanics and developed his defense of Copernican astronomy. He lectured to large audiences; and princes and nobles came to study under him.

When a bright new star suddenly appeared in the sky, he gave three lectures on it. Crowds came to hear him, but he rebuked them for paying attention to a temporary phenomenon while they overlooked the wonders of everyday nature. His lectures became so popular that even the great hall of the school of medicine was sometimes too small and he lectured in the open air. He taught honest science with compelling force.

Copernican Astronomy

Early in his career Galileo was converted to the Copernican system, and he taught it quietly at first, incautiously later. In a dialogue he describes what was probably his own conversion:

"I must upon this occasion relate some accidents that befell me when I first began to hear of this new doctrine [the Copernican system]. Being very young, and having scarcely finished my course of philosophy, . . . there chanced to come into these parts . . . a follower of Copernicus, who in an Academy made two or three lectures upon this point, to whom many flocked as auditors, but I, thinking they went more for the novelty of the subject than otherwise, did not go to hear him, for I had concluded with myself that that opinion could be no other than a solemn madness. Questioning some of those who had been there, I perceived they all made a jest of it, except one. He told me that the business was not altogether to be laughed at, and, because this man was reputed to be very intelligent and wary, I repented that I was not there. From that time forward, as often as I met with anyone of the Copernican persuasion, I demanded of them if they had been always of the same judgment; and, of as many as I examined, I found not so much as one who did not tell me that he had been a long time of the contrary opinion but had changed it for this, as convinced by the strength of the reasons in its favour. Afterwards, questioning them one by one, to see whether they were well possessed of the reasons of the other side, I found them all to be very ready and perfect in them; so that I could not truly say that they had taken up this opinion out of ignorance, or vanity, or to show the acuteness of their wits.

"On the contrary, of as many of the [Aristotelians] and Ptolemeans as I have asked (and out of curiosity I have talked with many) what pains they had taken in the book of Copernicus, I found very few that had so much as superficially perused it. But of those whom I thought had understood it, not one; and, moreover, I have enquired amongst the followers of the [Aristotelian] doctrine if ever any of them had held the contrary opinion and likewise found none that had. In other words, there was no man who followed the opinion of Copernicus who had not been first on the contrary side and who was not very well acquainted with the reasons of Aristotle and Ptolemy; and, on the contrary, there is not one of the followers of Ptolemy who had ever been of the judgment of Copernicus and had left that to embrace this of Aristotle. Considering these things, I began to think that a man who leaves the opinion imbued with his milk and followed by very many to take up another owned by very few and denied by all the schools, one that really seems a very great paradox, must needs have been moved, not to say forced, by more powerful reasons. For this cause I am become very curious to dive into the bottom of this business."[7]

Galileo's Mechanics of Motion

At Pisa and Padua, Galileo collected his knowledge of mechanics that he set forth much later in *Two New Sciences*. One of his early discoveries was the remarkable property of pendulums: that (for small amplitude) the time of swing is independent of amplitude. There is a fable that he discovered this in his student days in Pisa by timing the decreasing swings of a long lamp hanging in the cathedral. He had no accurate clock—in fact he was discovering the basis for good clocks—so he used his own pulse for the timing.[8] He then turned his discovery to use in medicine by constructing a short adjustable pendulum for timing pulses.

At some early stage, Galileo studied falling bodies, and he knew it was nonsense to say that heavier bodies fall faster *in proportion to their weight*. This had come from Aristotle—who probably gave it as a sensible description of final velocity in a very long fall, when air friction has increased

---

[7] Reprinted from the Salusbury translation of Galileo's *Dialogue*, as revised by Giorgio de Santillana, by permission of the University of Chicago Press. Copyright, 1953 by The University of Chicago Press. All rights reserved. Pages 142-144.

[8] The lamp quoted in the fable was not installed till some years later, so the story is doubted. Yet Galileo wrote in a dialogue ". . . Thousands of times I have observed vibrations, especially in churches where lamps, suspended by long cords, had been inadvertently set in motion."

till it balances weight. However, this was being taught as a rule for simple fall from rest—a piece of nonsense made almost sacred by ages of dogmatic teaching. Galileo saw that unequal bodies fall together, with motion *independent* of their weight, except for relatively slight differences. He satisfied himself by experiment and argument that those differences are due to air resistance. He pointed out that though a piece of gold falls fast, the same piece when beaten out to a thin leaf flutters down slowly. And he suggested a clinching proof: drop a scrap of lead and a wisp of wool in a vacuum—impossible in his day, but later carried out by Newton. He complained bitterly at the Aristotelians who claimed that in the time taken by a 100-pound cannon ball to fall 100 feet a 1-pound ball would fall only 1 foot. Actual experiment, he said, showed only a difference of finger-breadths. "How can you hide Aristotle's 99-foot difference behind a couple of finger-breadths?" he asked, to ridicule his opponents.

Galileo confirmed his belief of equal fall by comparing pendulums with light and heavy bobs on equal threads. The time of swing is the same, whatever the bob. Here was gravity-fall diluted, in a form that was easily timed accurately (a bunch of swings), and practically free from friction-troubles. (Since the time-of-swing is independent of amplitude, *air friction should not affect the time of swing*. Friction does reduce the amplitude from one swing to the next, but that does not matter!) This result contributed to the ideas of mass and gravitation that Newton later extracted and used. A heavy body has more weight than a light one—the Earth pulls it more. On that score we should expect it to fall faster. However, it also has more stuff-in-it-to-be-moved than a light one, a greater quantity of matter or mass, as Newton later called it. It has more "inertia," needs more force for its acceleration.[9] Therefore, when experiment shows that heavy and light bodies fall with the same acceleration (or swing with the same motion as pendulums) it suggests that the heavier body has greater mass in just the same proportion as it has greater weight. This is a remarkable property of gravitation, that Earth-pulls are proportional to the *inertial* masses of matter

[9] Though mass is an idea partly invented and partly extracted from properties of nature, it is not something that can be described successfully in a few words. One has to gain a feeling of its nature by working with it—calling it "inertia" is mere naming. When Newton defined it as "quantity of matter" he simply moved the doubt on to the definition of matter. Yet this description of Newton's was not as worthless as some critics claim. As a descriptive phrase it helped the scientists of Newton's day to understand what he meant. Perhaps it is useful in the same way for students today.

pulled. Galileo seems to have accepted it without seeking its cause. He did not formulate the concept of mass clearly in his studies of force and motion. It was Newton who put it to use. In this century mass gained new importance when we came to think of it as specially related to energy.

To investigate the motion of a falling body in detail, Galileo diluted gravity by using an inclined plane. He describes an experiment of rolling a ball

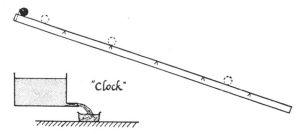

FIG. 19-1. GALILEO'S EXPERIMENT

down a long shallow incline, a grooved plank with a smooth parchment lining. Time-of-travel was measured with a simple water clock: the experimenter had a tank of water with a small spout and weighed the water that ran out. Measurements of time and distance agreed with the simple relation:

DISTANCE TRAVELLED FROM REST $\propto$ TIME^2

It is not clear whether Galileo actually did the experiment or just quoted it from earlier scientists. Anyway, the measurements were rough, but Galileo was confident he knew the correct "law." By an ingenious geometrical argument, he proved that this is the necessary rule for motion with constant $\Delta v/\Delta t$. Therefore the rolling ball moved with constant acceleration. By confident extrapolation from his shallow incline to a steeper one to a still steeper one and finally to vertical fall, he argued that freely falling bodies have constant acceleration; hence he knew their law of fall.

On any chosen incline, the force producing the acceleration must be the same all the way down. (It is a constant fraction of the ball's *weight*.) Already part of Newton's Law II had emerged: a constant force leads to constant acceleration.

Continuing with hills of different slopes, Galileo was on the verge of finding the main relation of Newton's Law II: ACCELERATION varies as FORCE; but he kept this in geometrical forms which obscured the part played by force. He was preparing the way for an experimental science of motion which could be applied to a variety of problems: projectiles, pendulums, the planets themselves; and, later, moving machinery and even the moving parts of atoms.

88

Speed at Bottom of Hill

Galileo guessed that if a ball rolls down one incline, A, and up another, B, *it will roll up to the original level*, whatever the slopes; and that led him to a very important general assumption from which he made many predictions. Imagine several downhill slopes A_1, A_2, A_3, *all of the same height*, all leading to the same uphill slope B. Then if his guess were true, the ball would rise to the same height on B *whichever of the A slopes it descended*. At the bottom, just about to run up B, the ball has the momentum needed to carry it up to the final point on B. That momentum must therefore be the same at the bottom of hill A_1, hill A_2, . . . ; the same for all slopes. Therefore the ball must have the same speed at the bottom of A, whatever the

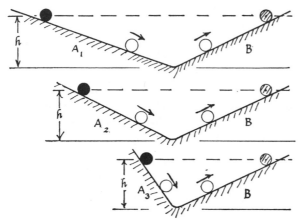

Fig. 19-2a. Galileo's Essential Fact
Ideal downhill-and-uphill motion.

slope. So Galileo made his general assumption: *The speeds acquired by any body moving down planes of different slopes are equal if the heights are equal.* This is the property sketched in Ch. 7, where we showed it belongs with Newton's Law II. Galileo generalized this to curved hills. He deduced many geometrical predictions for motion down inclines from this assumption, combined with his proven knowledge of constant acceleration.

The Downhill = Uphill Guess

Friction would prevent a satisfactory demonstration of a ball rolling down one hill and up another to the same height. Galileo probably based his guess on a mixture of experiment and thinking—he had a genius for making the right intuitive guesses with the help of rough experiments. It seemed plausible. For his colleagues he made it more plausible by a careful argument about compounding motions downhill and uphill. Look at the following irritating Galilean "thought experiment" (due to a later au-

thor): Suppose the ball finished *higher* on the further slope B. We could insert an extra plank, C, and let it roll back to its starting-place on A, and then start again with the velocity it had gained. This could continue, with the ball gaining more and more motion, cycle after cycle, which seems absurd. If the ball finished *lower* on the further slope B, then we might start it on B instead, at that lower point. If it rolled down B we should expect it to retrace its path to the original point on A, thus ending up higher in this backward journey. Again with an extra plank, D, we could arrange for cycle after cycle of an absurd increasing motion. Both cases seem absurd and therefore the ball must rise to the same height

Fig. 19-3. A Galilean "Thought Experiment"

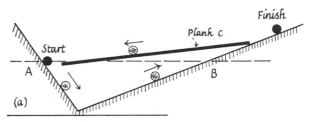

Fig. 19-3a. If the ball finishes higher, let it run back down a temporary extra plank, C. Absurd increasing motion.

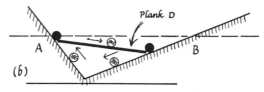

Fig. 19-3b. If the ball finishes lower, let it make the return trip, presumably (?) retracing its path and ending higher on A than on B. Then let it roll back, via a temporary plank D, to the same place on B. Absurd, perpetually increasing motion.

on the opposite hill. The weak point in the argument is the claim in the second case that the motion must be exactly reversible. Even apart from this, the method of solving problems by argument does not seem to us very scientific. But Galileo lived in an age of arguers and knew that such attacks would carry considerable weight. Besides, here, as elsewhere in physics, arguments *can* help to clarify the problem, to suggest what to think out and what to investigate.

Galileo himself clinched his contention by producing what seems impossible, a frictionless version of the downhill = uphill experiment, an amazingly simple but convincing demonstration, his "pin and pendulum" experiment. In this a peg catches the thread of a pendulum as it swings through its lowest point, thus converting the pendulum abruptly from a long one to a short one. In all cases, the bob after

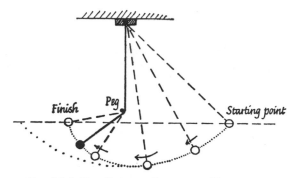

FIG. 19-4. THE PIN AND PENDULUM EXPERIMENT
The pendulum bob rises to the same vertical level, in spite of the change of motion caused by the peg.

falling down a long shallow arc rises up a steep arc to the same height. Try this experiment of genius yourself.

In this property of nature, Galileo had the key to one aspect of conservation of energy, though that general idea was not formulated till later.

Newton's First Law of Motion

The downhill=uphill rule suggested one thing more. Galileo argued to the extreme case when the second hill is horizontal, with no slope. Then a ball that rolls down the first hill must continue along the horizontal forever. Thus he found the essence of Newton's First Law of Motion: *Every body continues to move with constant velocity (in a straight line), unless acted on by a resultant force.* Here, had he but known, he had the key to one puzzle of planetary motion: What forces maintain the motion of the planets, Moon, etc.? What pushes them *along* their orbits? The new answer was to be: no force; none is needed, because motion continues of its own accord.

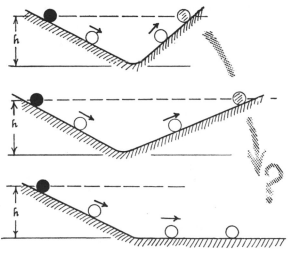

FIG. 19-2b. GALILEO'S ARGUMENT

Independence of Motion

The parallelogram addition of velocities, forces, etc., was just being recognized or discovered—essentially this implies that one vector does not disturb another: they act independently and just add geometrically. All through his experiments Galileo insisted that motions (and forces) are independent of each other. For example, a vertical accelerated motion and a constant horizontal motion simply add by vectors—one motion does not modify the other, but each has its full effect. He applied this to ideal projectiles and showed that their paths are parabolas.

Galileo preached this independence of vectors again and again in his dialogues, as an essential reply to critics of Copernicus. When they claimed that a moving Earth would leave falling bodies far behind, Galileo asked about things dropped from the mast of a ship that was sailing steadily. If they then murmured about wind, he repeated the "thought experiment" in the ship's cabin. He said that clouds and air, which already have the motion of the Earth's surface, simply continue to move with it. He carried his readers through problems like those at the end of Chapters 1 and 2 and showed that a steady motion of the laboratory does not affect experiments on statics, free fall, or projectiles. A laboratory's steady motion cannot be detected by any mechanical experiments inside. That is *Galilean Relativity.*

Leaving Padua

After 20 years at Padua, Galileo was tempted to return to his home university at Pisa. He had kept in touch with the Medici family there, and now he negotiated with the Duke for an appointment in Florence with better pay and more leisure. His public duties at Padua took only an hour a week, but to supplement his salary—which was still small, though an admiring university increased it several times—he had to do private teaching. "He was weary of universities, of lecturing, tutoring, and boarding students; he had had enough of the stuffed robes against which he had written satirical poems; . . . of the closed and petty atmosphere of Padua. . . . He wanted to be in his own land, in his own native light and air, free, and among friends of his own choosing."[10] He needed leisure to study and write, and the support of noble patronage. In return

[10] G. de Santillana in his "Historical Introduction" to Galileo's *Dialogue, op.cit.,* p. xi.

for a better salary he promised the Duke he would write a series of books: ". . . principally two books on the System of the Universe, an immense design full of Philosophy, Astronomy, and Geometry; then three books on Motion, three on Statics, two on the Demonstration of Principles, one of Problems; also treatises on Sound and Speech, Light and Colors, Tides, the Composition of Continuous Quantity,[11] the Motion of Animals, and the Military Art." This gives some indication of his eagerness and wide interests.

The Telescope

While he was considering the move to Pisa and Florence, Galileo happened to hear of the invention of the telescope. It is said that a Dutch spectacle-maker[12] had found an arrangement of two lenses that made distant things seem large and close. Hearing about this, Galileo made a simple telescope, magnifying only three times, by fixing two lenses in a pipe. He used a weak convex lens for the first lens, and a concave lens for the eyepiece. This may have differed from the invention he heard of, in which case he was the first to make an "opera glass" which we now sometimes call a Galilean telescope. He was delighted with his new instrument and the fame of this marvel soon spread. The telescope was the talk of society, and crowds came to look through it. The Venetian Senate hinted they would like a

[11] The *composition of continuous quantity* sounds like an attempt at the calculus problem of integration. The need for calculus as a mathematical tool was growing. Galileo himself needed it, and made preliminary attempts. By the next generation the time was ripe for its development, and it is not surprising that Newton and Leibniz invented it independently.

[12] The spread of printing in the century before had increased the use of spectacles, so the time was ripe for the discovery of telescopes by the chance putting together of lenses.

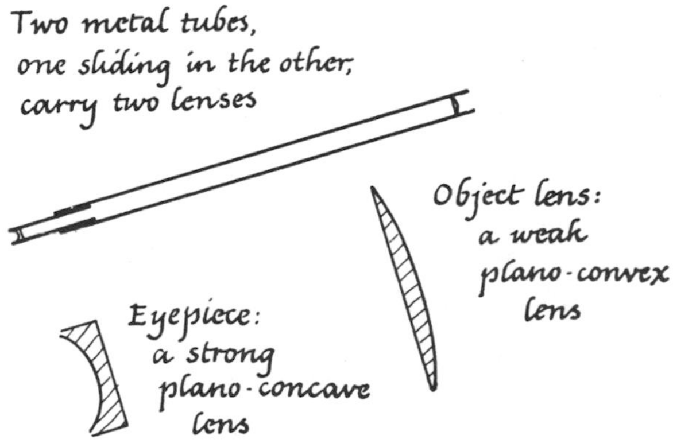

Two metal tubes, one sliding in the other, carry two lenses

Object lens: a weak plano-convex lens

Eyepiece: a strong plano-concave lens

Fig. 19-5. Galileo's Telescope

copy. Galileo presented them with one. His salary was doubled soon after!

Galileo looked at the Moon and then the planets and stars with, he says, "incredible delight." If you have never seen the Moon through a telescope, borrow any small instrument, field-glasses or a toy, and try it.

On observing the Moon, Galileo saw mountains and craters. He even estimated the heights of Moon-mountains from their shadows. What he saw was unwelcome to many who had been taught that the Moon is a smooth round ball. Mountains and craters made the Moon earthy and broke the Aristotelians' sharp distinction between the rough, corruptible Earth and the polished unchangeable heavens. The telescope dealt a smashing blow to the old astronomy of perfect spheres and globes. Human beings are conservative and do not like to have their settled opinions changed by a newcomer who proves he is right. Far from pleased at being shown something new, they are angry to find their beliefs upset, particularly if those beliefs have been firmly established in childhood—their sense of security is assailed. So Galileo found some people angry over his discovery. When he offered a convincing look through his telescope, many were delighted, but some refused, and others looked and then said they didn't believe it. One Aristotelian admitted the mountains were there but explained away the damage by saying that the valleys between them are filled with invisible crystal material to bring the surface back to a perfect sphere. Sure, said Galileo, and there are mountains of invisible crystal there as well, that stick out ten times as far!

Good lenses were hard to obtain and Galileo had to grind and polish his own in his instrument shop. He made better ones than most, so that his telescopes succeeded where others failed. Even so, his telescope (preserved in Florence) gives a poor image compared with modern instruments. He made a second instrument which magnified eight times, then another which magnified 30 times and which involved great labor: the grinding of a block of glass into the right shape is a tedious, difficult business, and much of the final performance depends on care in polishing. Through the new telescope the planets appeared as bright discs. The stars looked brighter and farther apart, but were still just points. Galileo was delighted to find how many more stars he could see. The luminous haze of the "milky way" was resolved into a myriad of stars.

FIG. 19-6a.
A PHOTOGRAPH OF THE MOON NEAR LAST QUARTER (Photograph by J. H. Moore and J. F. Chappell with 36-inch refractor, at Lick Observatory.) The sunshine catching the mountain tops near the edge shows the rough landscape that Galileo's telescope revealed, to the dismay of many people. [This shows the Moon inverted, as we see it with a modern telescope.]

Jupiter's Moons

With this new powerful telescope Galileo made a still more important discovery. On the night of January 7, 1610, he observed three small stars in a line (Fig. 19-7) near Jupiter, two to the East of Jupiter, and one to the West. He thought they were fixed stars and paid little attention to them. The next night he happened to look at Jupiter again and found all three of the stars were to the West of Jupiter and nearer one another than before. He ignored the latter peculiarity and thought the shift was due to Jupiter's motion; but then he realized that this would require Jupiter to have moved in the *wrong direction*, for Jupiter was on a backward loop. This was mysterious indeed. He waited anxiously to observe them again the next night but the sky was cloudy. The night after that, on January 10, only two of the stars appeared, both to the East

FIG. 19-6b. A PHOTOGRAPH OF MOON MOUNTAINS (Lick Observatory). This is a section of the picture (a), enlarged about 6 times. The peak with the long shadow near the bottom is the mountain "Piton."

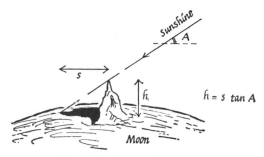

FIG. 19-6c. SKETCH OF PITON AND ITS SHADOW, to show how mountain heights are estimated. (After Whipple, in *Earth, Moon and Planets.* Harvard University Press, 1941.)

of Jupiter. Jupiter could hardly have moved from West to East in one day and then from East to West in two days by such amounts. Galileo decided the "stars" themselves must be moving and he set himself to watch them. The sketches show his record of what he saw. He had really discovered four small moons moving around Jupiter.

Look at Jupiter yourself, with any small telescope, even field glasses. You cannot miss the moons, which you will see better than Galileo did with his simple lenses.

Kepler, when Galileo wrote to him about his discovery, shared Galileo's delight, although the extra moons seemed rather contrary to his limit of six

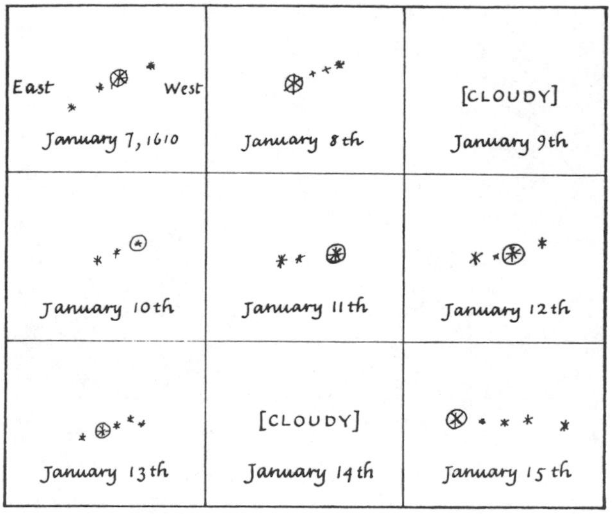

FIG. 19-7. GALILEO'S OBSERVATIONS OF JUPITER'S MOONS These sketches are copied from Galileo's handwritten record. (The orbits of the moons are nearly in planes containing our line of sight from Earth to Jupiter; so the moons are often in front of Jupiter or behind, and they are often eclipsed by moving into Jupiter's shadow. They move quickly around their orbits. That is why the pattern changes so quickly and why, often, less than four moons are visible.) (For a copy of Galileo's written record, see *Galileo* by J. J. Fahie.)

planets. The Aristotelian philosophers did not welcome the discovery because it made Jupiter an important center, spoiling the Earth's unique position, supporting, in fact, the Copernican theory. One argued,

"There are seven windows in the head, two nostrils, two eyes, two ears, and a mouth; so in the heavens there are two favorable stars, two unpropitious, two luminaries, and Mercury alone undecided and indifferent. From which and many other similar phenomena of nature, such as the seven metals, etc., which it were tedious to enumerate, we gather that the number of planets is necessarily seven. Moreover the satellites of Jupiter are invisible to the naked eye, and therefore can have no influence on the earth, and therefore would be useless, and therefore cannot exist. . . ."

Galileo wrote to Kepler:

"Oh, my dear Kepler, how I wish that we could have one hearty laugh together! Here, at Padua, is the principal professor of philosophy whom I have repeatedly and urgently requested to look at the moon and planets through my glass, which he pertinaciously refuses to do. Why are you not here? What shouts of laughter we should have at this glorious folly! And to hear the professor of philosophy at Pisa labouring before the grand duke with logical arguments, as if with magical incantations, to charm the new planets out of the sky."

Jupiter and his moons provided a small scale model of the Copernican system of Sun and planets —a compelling argument for the Copernican view. Kepler used Galileo's measurements in a rough test to see whether his Law III applied to Jupiter's moons. He found it does apply, though the constant obtained for R^3/T^2 is different from the value given by the Sun's planets. The table below shows modern, more accurate, data. More moons, smaller and farther out, have been discovered since. We now know a dozen.

Return to Pisa and Florence

When Galileo accepted the new post at Florence he had to resign his professorship at Padua. This unexpected resignation was most unwelcome there; it seemed ungrateful and even unfair, but the new post offered better opportunities. Galileo made the move and lost some friends. Though he now enjoyed the leisure he needed for his work, the move proved unwise in the long run, because it was a return to enemies as well as friends. (Back in his student days there he had been known as "the

JUPITER'S SATELLITES AND KEPLER'S THIRD LAW

NAME OF SATELLITE	DISTANCE FROM JUPITER in Jovian diameters	in miles (R)	TIME OF REVOLUTION in hours (T)	CALCULATIONS FOR TEST OF LAW III R^3 (miles)^3	T^2 (hours)^2	$\dfrac{R^3}{T^2}$
Io	3.02	262,220	42.36	1.803×10^{16}	1802.8	TRY
Europa	4.80	417,190	85.23	7.261×10^{16}	7264.	THIS
Ganymede	7.66	665,490	171.71	29.473×10^{16}	29,484.	(See
Callisto	13.48	1,170,700	400.54	160.440×10^{16}	160,430.	note 14)
	(see note 13)					

[13] It is simplest to measure the moons' orbits in terms of Jupiter's diameter. The radii could remain in those units for a test of Kepler's Law III; but, if these data are to be used in gravitational theory (e.g., to compare Jupiter's mass with the Sun's), then the same units, e.g., miles, must be used on both sides of the comparison.

[14] The test is made easy by a lucky chance arising from the choice of units, miles and hours. Look at the numbers.

wrangler," and he had been violent in his attacks on those he called the "paper philosophers.") He was a sincere man but not tactful, and the opposition which his discoveries and arguments excited was more a subject of triumph to him than sorrow. "The Aristotelian professors, the temporizing Jesuits, the political churchmen, and that timid but respectable body who at all times dread innovation, whether it be in religion or science, entered into an alliance against the philosophical tyrant who threatened them with the penalties of knowledge."[15]

At Florence he continued to study the planets with his new telescope and soon made more discoveries. He found that Saturn seemed to have a small knob on each side, almost fixed to it. Modern telescopes show Saturn as a bright ball with a flat ring round it like a hat-brim; and we now know that the ring is made up of small pebbles, possibly ice, all circulating independently around Saturn—an army of Kepler III examples. In Galileo's telescope the ring was not clear—seen almost edge-on, it looked like side-planets. Galileo next found that Venus showed phases like the Moon. Here was direct confirmation of the Copernican picture. On the Sun, Galileo found sunspots, black blotches that moved and changed—another blow at the purity of the heavens.

He took his telescope on a tremendously successful visit to Rome, where he was welcomed with enthusiasm. His telescope was the wonder of the day, and a Church committee approved his discoveries.

Growing Troubles

Galileo returned to Pisa full of plans for a great treatise on the Constitution of the Universe. The Copernican system seemed right in its simplicity and compellingly vouched for by his telescope. An Earth spinning with its own momentum would remove the improbable daily motion of an outermost sphere of stars, driven by "an immense transmission belt from nowhere,"[16] with uncouth gears to run the inner spheres. A central Sun, with Earth a planet, simplified many heavenly motions and predicted what he saw in his telescope. He even saw a model of the Solar System in Jupiter and his moons. But that Copernican view conflicted with the simple poetic astronomy of the Bible, taught with full authority of the Churches, Roman Catholic and Protestant alike. Just as Galileo felt confident he could prove the Copernican case, Church disapproval suddenly hardened. He was attacked in sermons,

and his arguments in reply were sent secretly to the Inquisition in Rome. From Rome, friends warned him the Copernican doctrines were under grave question. Galileo's pupils and friends, including the Duke himself, defended him nobly, but he started his most serious troubles by writing dangerous public letters on scripture and science. In these he claimed that the language of the Bible should be taken metaphorically and not literally, where science is discussed. The Bible, he said, teaches us spiritual matters, but not the facts of nature—and he quoted a cardinal: "The Holy Spirit teaches us how to go to Heaven, not how the heavens go." Since scripture and nature are both works of the same divine author, they cannot really be in conflict, but they serve different purposes; and the Church should not try to make astronomers disbelieve what they see. Nor should people condemn Copernicus' book without first reading and understanding it. Such was his open defense.

Authorities in Rome were still more disturbed. Church astronomers withdrew their support of Galileo in the light of this subversive attack. Galileo, alarmed, went to Rome to investigate his own safety. There he argued persistently with friend and foe alike; but "to such minds, Galileo could not communicate what he and Kepler were alone to see: the three forces of mathematics, physics, and astronomy converging toward a junction which would make them irresistible and creating a physical science of the heavens."[16] Meanwhile the Church had appointed a group of theological experts to examine the Copernican teaching and they reported on two key propositions:

THAT THE SUN DOES NOT MOVE: ". . . false and absurd in philosophy, and formally heretical."

THAT THE EARTH BOTH MOVES AND SPINS: ". . . false and absurd, and at least erroneous in faith."

Galileo stayed on in Rome to help the discussion, as he thought. He was summoned and told that the Copernican doctrine was condemned as "erroneous." Copernicus' book was suspended—no devout Catholic could read it until it had been "corrected." And Galileo himself was not to hold or defend the doctrine as true. He waited a short while, to show a brave face, then returned home, in good standing as a devout Catholic, but reproved and bitterly disappointed.

He remained in Florence for half a dozen years. A new Pope was elected to the throne who was friendlier to science and was in fact a friend of

[15] Sir David Brewster, *Martyrs of Science* (1848).
[16] G. de Santillana, *op.cit.*

Galileo. Delighted, but in poor health, Galileo made an uncomfortable journey to Rome to congratulate the new Pope, and he had a marvelous visit there. He had several audiences with the Pope, who gave him rich presents and honors. He even argued delicately on the Copernican system, emphasizing its simplicity. The cardinals were reserved, but the Pope himself commented, "The Church has not condemned this system. It should not be condemned as heretical but only as rash." However, when Galileo pressed his views, the Pope replied sharply that the earlier prohibition must stand. He told Galileo not to limit the wisdom of God to a scientific scheme: *God could devise any scheme he pleased*—a very able argument that can stop all science. However, the Pope finally agreed that Galileo might write a *non-committal* book explaining the arguments on both sides between Copernicus and Ptolemy. That would be a mere theoretical discussion, leaving any question of fact and truth to be decided by the higher wisdom of the Church.

The Great Dialogue

Galileo returned home, disappointed yet honored and confident. He was confident that he had permission to write his long-planned book on the System of the Universe. But he was overconfident, and perhaps undergrateful to the Church. He continued surreptitious teaching of Copernican ideas and developed his book. He wrote it in the form of a dialogue—a very acceptable form of teaching in those days. After some difficulty with Church censors, one of them a personal friend, Galileo got the book published. Its title runs

THE DIALOGUE

OF

GALILEO GALILEI, Member of the Academy of Lincei

PROFESSOR OF MATHEMATICS IN THE UNIVERSITY OF PISA

And Philosopher and Principal Mathematician to

THE MOST SERENE

GRAND DUKE OF TUSCANY

Where he discusses, in four days of discourse,

the two

GREAT SYSTEMS OF THE WORLD
THE PTOLEMAIC AND THE COPERNICAN

Propounding impartially and indefinitely the Philosophical and Physical arguments, equally for one side and for the other side

It begins with a preface addressed "to the prudent reader" which looks like a most imprudent attack on the Inquisition. The dialogue is conducted by Salviati, a philosopher, who sets forth with able arguments Galileo's Copernican views; Sagredo, who, as a sort of attorney to Salviati, asks questions and raises difficulties and cheers up the dialogue with his wit; and Simplicio, a dogged follower of Aristotle and Ptolemy, who is beaten in argument by Salviati every time and made a fool of by Sagredo.

The *Dialogue* was not written in Latin for scholars but in Italian, for the general reader, in rolling prose with long discussions and cunning arguments. One critic says it "meanders at ease across the whole cultural landscape of the time";[17] but essentially it was a great setting forth of the nature of motion, terrestrial and celestial, with the fullest arguments in favor of the Copernican scheme. It had much good teaching and argument, and some serious shortcomings. Galileo was a great man, but the new science that he taught was still unfinished and sometimes unclear. He never realized that circular orbits need an inward pull. He maintained *vertical fall* as natural for *earthly* bodies and *circular motion* as natural for *celestial* bodies—an Aristotelian prejudice—despite his understanding of inertia in earthly mechanics. He never taught Kepler's elliptical orbits, perhaps because of that principle of circles, or perhaps because as a hard-headed teacher he realized how *very* close to circles the actual ellipses are. His explanation of tides—caused by a breathing Earth—seems even more pig-headed.

The book proved popular and carried conviction. By contrast, Copernicus' book was difficult—few had understood its full import, and now it was prohibited. There had been talk and covert discussion, but most educated people could not "piece together the great puzzle that stayed disassembled by superior orders. The *Dialogue* did exactly that: it assembled the puzzle and for the first time showed the picture. It did not go into technical developments; it left all sorts of loose ends and hazardous suggestions showing to the technical critic. But it was exactly on the level of educated public opinion, and it was able to carry it irresistibly. It was a charge of dynamite planted by an expert engineer."[18]

News of this tremendous attack soon reached Rome and the Pope, good friend though he was, ordered the Inquisition to forbid the book and re-examine Galileo. Galileo, aged and sick, was summoned to

[17] G. de Santillana, "Historical Introduction" to Galileo's *Dialogue, op.cit.*, p. xxx.
[18] G. de Santillana, *op.cit.*, p. xxxi.

Rome. There he was well treated on the whole and and comfortably housed—they knew he was a great man—but the Inquisition proceeded to a strict examination, formulating his offenses then asking him to defend his actions. Galileo knew he had written dangerously, but with permission. The original prohibition had only instructed him not to teach Copernican astronomy as *true*, and he had obeyed the dictum, thinly and insincerely. So he felt fairly safe until a document, probably forged, was produced that showed he had promised never to teach or discuss the Copernican system at all. In that case, the indictment was very serious: heretical teaching and writing in the face of a pledge to desist. Failing repentance and abjuration, a heretic faced terrifying threats of torture, and then torture. Galileo was in a very grave position. He had disobeyed instructions of the Church; he had set forth the Copernican picture *in print* (with a thin pretense that it was only a piece of theorizing); and he had even criticized the interpretation of the Scriptures. The powerful, ruthless Church, which crushed questioners with severe punishment and condemned defiant martyrs to be burned at the stake, would stand no such behavior. Outside the Inquisition's court, he was well treated; inside, they were lenient at first, reasoning with him and asking him to defend his position. Yet he was being questioned by a court that held both the physical powers of torture and the spiritual powers of a great Church. His health grew worse; he was questioned and questioned again. Still he held on to his beliefs, holding on to his real life. A friendly examiner suggested he should confess to false pride as the cause of his writing and be let off lightly. Galileo, giving up hope of arguing his position, at last agreed. However, the highest court of the Church overruled this lenient compromise and insisted on unconditional surrender. Galileo was summoned to a "rigorous inquiry." He did not emerge from the court till three days later. We do not know how far he was taken in the steps towards torture. He was not tortured physically— that was ruled out by his great age—yet to his intellect much of the proceedings must have had the horrors of mental torture. In the course of this inquiry he agreed to recant completely, to withdraw his unorthodox statements and deny his own earlier beliefs. He accepted the judgment of the Inquisition as a penitent—remember he was a pious if argumentative member of the Church—and he knelt and read the abjuration required of him, swearing never again to believe in or teach the Copernican system. It was a long grim document of abject apology, con-

fession of errors, complete recantation of views, and absolute promises for the future under severest penalties. Kneeling, he signed.

There is a tale that as he rose from his knees he muttered "E pur si muove"—"and yet it [the Earth] does move"—but that is hardly likely. There was no friend there to hear, it was far too dangerous, and Galileo was a broken old man. As Bertrand Russell has said, "It was the world that said this—not Galileo."

Galileo was imprisoned for a time, then allowed to return home, under some restrictions. His health was poor, but his head, he complained, was "too busy for his body." He composed his great discourse on *Two New Sciences*. It contained the account of his work on accelerated motion which formed the basis for Newton's laws, his rules for elasticity of beams, and his foundations for calculus. This was no popular account, but a great technical text. While he was working on it he became blind in one eye and soon the other one became blind too. He says of this calamity: "Alas! your dear friend and servant has become totally and irreparably blind. These heavens, this earth, this universe, which by wonderful observation I had enlarged a thousand times beyond the belief of past ages, are henceforth shrunk into the narrow space which I myself occupy. So it pleases God; it shall therefore please me also." He was now allowed more freedom, and in spite of much sickness, he continued his writing with the help of friends. His health grew worse, and he died at the age of 78.

The Contest between Science and Church

Galileo brought into the open the differences between authoritarian churchmen and independent scientists. By his tactless manner and powerful arguments he brought great troubles on himself, and on science too. His biographers differ in their views of his conflict with the Roman Catholic Church, according to their own feeling about authority. Some paint him as almost a martyr, threatened with torture by a bigoted Inquisition, suspected, persecuted, imprisoned, and forbidden to teach the great Truths he had helped to discover—with the Church as the villain of the piece taking the side of superstition and prejudice, trying to suppress, in the interest of dogmatic authority, the simple things of nature that should be to the glory of a worldwide religion. Others show Galileo bringing his troubles on himself by his hotheaded arguments and exasperating manner of setting people right; they paint him as ungrateful towards the Church which listened to his teaching and honored him with pensions; and they point out that his conflict with the Church arose directly from his attack on scriptural science—in which he was meddling in the Church's rightful province. Others again regret his subservient behavior—in not making himself a martyr for science —but this seems a cruel criticism of one who was so far from subservient most of his life.

Bertrand Russell says,

"The conflict between Galileo and the Inquisition is not merely the conflict between free thought and bigotry or between science and religion; it is a conflict between the spirit of induction and the spirit of deduction. Those who believe in deduction as the method of arriving at knowledge are compelled to find their premises somewhere, usually in a sacred book. Deduction from inspired books is the method of arriving at truth employed by jurists, Christians, Mohammedans, and Communists. Since deduction as a means of obtaining knowledge collapses when doubt is thrown upon its premises, those who believe in deduction must necessarily be bitter against men who question the authority of the sacred books. Galileo questioned both Aristotle and the Scriptures, and thereby destroyed the whole edifice of mediaeval knowledge. His predecessors had known how the world was created, what was man's destiny, the deepest mysteries of metaphysics, and the hidden principles governing the behavior of bodies. Throughout the moral and material universe nothing was mysterious to them, nothing hidden, nothing incapable of exposition in orderly syllogisms. Compared with all this wealth, what was left to the followers of Galileo?—a law of falling bodies, the theory of the pendulum, and Kepler's ellipses. Can it be wondered at that the learned cried out at such a destruction of their hard-won wealth? As the rising sun scatters the multitude of stars, so Galileo's few proved truths banished the scintillating firmament of mediaeval certainties. . . . Knowledge, as opposed to fantasies of wish-fulfilment, is difficult to come by. A little contact with real knowledge makes fantasies less acceptable. As a matter of fact, knowledge is even harder to come by than Galileo supposed, and much that he believed was only approximate; but in the process of acquiring knowledge at once secure and general, Galileo took the first great step. He is, therefore, the father of modern times. Whatever we may like or dislike about the age in which we live, its increase of population, its improvement in health, its trains, motor-cars, radio, politics, and advertisements of soap—all emanate from Galileo. If the Inquisition could have caught him young, we might not now be enjoying the blessings of air-warfare and poisoned gas, nor on the other hand, the diminution of poverty and disease which is characteristic of our age."[19]

The principal mistake was the same one on both sides: failure to regard the Copernican scheme as a useful picture, an hypothesis or piece of "theory." Galileo and the Church argued whether it was *true*.

". . . it is entirely false to assert that the Church stopped the scientific experiments of Galileo. Galileo's difficulties with the Church had nothing to do with his experiments. They developed, apart from purely personal causes, out of his refusal to yield to the request that he treat the Copernican hypothesis as an hypothesis, which in the light of modern relativity was not an unreasonable request. There seem to be as many myths about Galileo as about any of the saints."[20]

Galileo himself, with his love of truth, might have decided more calmly to treat the Copernican view as an hypothesis—though in mechanics he did believe in absolute fixed space. He might well have defended the wisdom of such an open mind.

The new scientific temper was unwelcome to the academic philosophers, busy with cultured discourse—though their spirited Renaissance predecessors might have enjoyed it. The Church, busy maintaining political power and spiritual good, feared it. (Earlier in Galileo's lifetime, Bruno had been burned at the stake for heretical views, among them his use of Copernican astronomy. He had pointed out that, with the outermost "sphere" of stars at rest, the stars could be spread into infinite space beyond—myriad suns occupying Heaven. That dissolving of our universe's neat outer shell was a shocking novelty to the medieval mind.) Galileo's *Dialogue* was placed on the forbidden list, and his abjuration was read in churches and universities, "as a warning to others." The new Protestant Church was no less intolerant. With no papal authority, its leaders placed even greater importance on the literal truth of the Bible. Martin Luther had said of Copernicus, "the fool will overturn astronomy." In Italy and elsewhere, the new science was

[19] Bertrand Russell, *The Scientific Outlook* (W. W. Norton and Company, Inc., New York, 1931), pp. 33-34.

[20] Morris R. Cohen, *The Faith of a Liberal: Selected Essays*, in which there is a delightful essay, reviewing Galileo's book *Two New Sciences*. These three pages (417-419), are well worth reading. (Henry Holt and Company, New York, 1946.)

discouraged by authority and restricted by fear for another half century.

This struggle is not unique, an isolated battle of the past. There is some such struggle in every age. Conditions of living, interacting with men's knowledge and powers and beliefs, produce struggles between conservative elements—not always the same kind of people—and rebellious ones—not always with the same kind of causes. In each age, from Galileo's to now, the struggles of the day have usually seemed just as serious, making it just as dangerous to take, or even to discuss, the rebel's side. Often, a generation or two later, both sides regret the quarrels; yet mankind still learns sadly little from those regrets. Age after age, current quarrels seem as vital and as dangerous as their forgotten or regretted ancestors did in their own time.

Here is a translation of a note in Galileo's handwriting in the margin of his own copy of the *Dialogue*:

"In the matter of introducing novelties. And who can doubt that it will lead to the worst disorders when minds created free by God are compelled to submit slavishly to an outside will? When we are told to deny our senses and subject them to the whim of others? When people devoid of whatsoever competence are made judges over experts and are granted authority to treat them as they please? These are the novelties which are apt to bring about the ruin of commonwealths and the subversion of the state."[21]

Whatever we make of Galileo's life, his work remains, as a foundation of physics, a monument of method for all who came after, and a basis of knowledge on which the science of mechanics was already being built.

[21] From the English translation edited by G. de Santillana, preceding p. xi. Chicago University Press, 1953.

CHAPTER 9 · THE SEVENTEENTH CENTURY

〜〜〜〜〜〜〜〜〜〜〜〜〜〜〜〜〜〜〜〜〜〜〜〜〜〜〜〜〜

"Great ideas emerge from the common cauldron of intellectual activity, and are rarely cooked up in private kettles from original recipes."

—James R. Newman
in *Scientific American*

〜〜〜〜〜〜〜〜〜〜〜〜〜〜〜〜〜〜〜〜〜〜〜〜〜〜〜〜〜

A Hundred Centuries of Astronomy

In the hundred or more centuries between the earliest civilized men and the time of Galileo, Astronomy grew from simple beginnings rooted in primitive man's curiosity and wonder and fear to a well-ordered science ready to furnish a great laboratory in which mechanics, the basis of all physics, could be developed and tested. For many centuries, astronomy remained in the hands of the calendar-makers and priests, with only a rare scientist to observe more carefully than the rest or to extract some new thread of law from the tangle of observation. Astrologers throve on men's superstitious fears; yet, just as alchemy did much for chemistry, astrology helped early astronomy. Then came philosophers and scientists with greater interest in knowledge for its own sake. They gathered careful observations and extracted rules for the motions of planets, Moon, and Sun—working rules, though clumsy ones. They invented reasons or causes which now seem fanciful and complex. Astronomy waited centuries more, almost asleep but for a few keen observers, while civilization reformed itself toward a new awakening. In those dark ages, the teaching of the Church and the method of deductive argument based on the authority of books were supreme in intellectual matters. Tradition replaced experiment, prejudice overrode science.

Yet there were practical needs, in medicine and navigation, to keep good science alive. Then came the cry with growing fervor, "Watch what *does* happen; stop arguing about what *ought* to happen." Prejudice was being pushed aside by careful thinking in terms of experimental observations.

The Renaissance

In the three centuries preceding 1600, the Renaissance grew and spread across Europe, a great awakening of new interests in art and literature and a new outlook in religion. The tight hold of the traditional scholastics loosened, and the study of Greek writers in the original opened out after centuries of obscurity; paper and printing came to spread knowledge, and the renewed interest in knowledge, far and wide; great navigations brought new markets, new wealth, new outlooks, and new leisure for intellectual growth; the arts flourished more freely and helped to release man's restless spirit of inquiry.

The Renaissance brought a new attitude of mind: the idea of *man as an individual* developed, in contrast with man as a servant to his group and its traditions. The enthusiasm for study and the general spirit of free inquiry that came with the renewed "humanities" prepared the ground for the development of science in the 17th century: ". . . the humanists . . . played the chief part in that widening of the mental horizon which alone made science possible. Without them, men with scientific minds would never have thrown off the intellectual fetters of theological preconception; without them, external obstacles might have proved insurmountable."[1]

In Renaissance science, one man towers above the rest, Leonardo da Vinci, a new scientist a century or two before his time. A genius at whatever he turned his hand and mind to—painting, sculpture, architecture, engineering, . . . physics, biology, . . . philosophy, . . . —he regarded observation and experiment as the only true approach to science. "He dismissed scornfully the follies of alchemy, astrology . . . ; to him nature is orderly, non-magical, subject to immutable necessity."[2] He trusted the logical conclusions of arithmetic and geometry because they are based on concepts of universal truth; but science must be based on experiment. He said, "those sciences are vain and full of errors which are not born from experiment, the mother of all certainty, and which do not end with one clear experiment."[3] His own experimenting with techniques for art and architecture and engineering led him to much scientific knowledge: machines and forces;

[1] Sir William Dampier, *A History of Science*, 4th edn. (Cambridge University Press, 1949), p. 98.
[2] *op.cit.*, p. 107.
[3] Quoted by Dampier, *op.cit.*, p. 105.

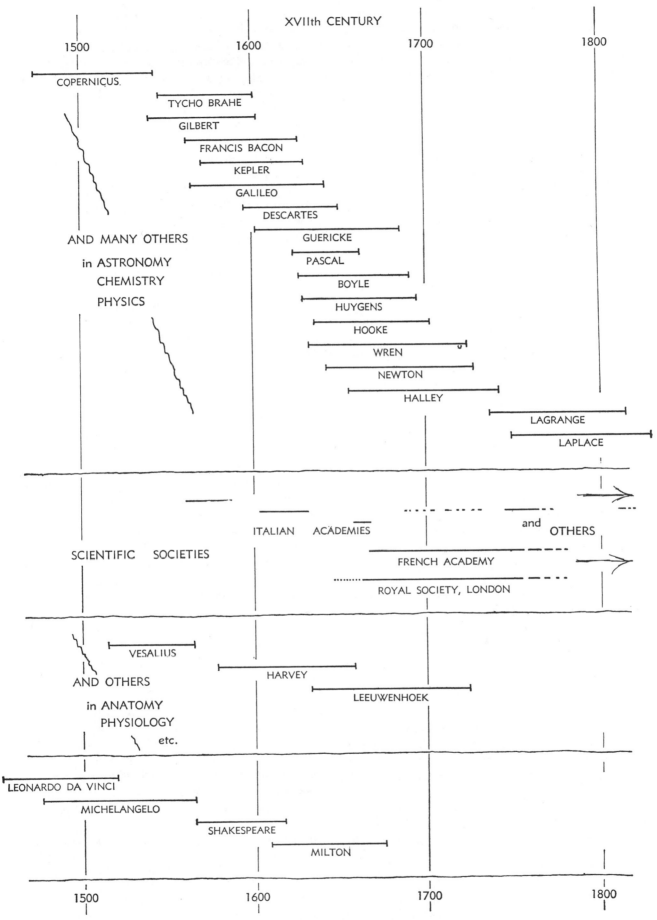

FIG. 20-1. A TIME CHART

properties of motion; Newton's Law I in simple form, long before Galileo taught it; properties of liquid pressure and fluid flow; waves in water and sound waves in air; the impossibility of perpetual motion. He advanced optics with studies of the eye and perspective; and he may have constructed a pendulum clock 200 years before Huygens. He saw in fossils the evidence of geological history; he may have discovered the circulation of the blood; and he set forth human anatomy in brilliant drawings that used the knowledge from many dissections. We learn all this from his drawings and from scrappy notes in his private notebooks. "Had he published his work, science must at one step have advanced to the place it reached a century later."[4]

"Progress"

Sometime then the idea of *progress* appeared, as a new outlook. We now take progress as an obvious aim—progress toward a better state, a higher education, a finer man, etc.—and we might think our ancestors always aspired to progress. Yet, for many centuries, as men looked back on a golden age in the past and tried to model their conduct on tradition, the idea of progress in earthly life had been almost unknown. When a forward outlook emerged from the Renaissance, it offered science new encouragement.

The Seventeenth Century

The simpler Sun-in-center idea suggested by Copernicus gained ground faster and faster, as new experimental evidence came to its support and freedom of speech and teaching increased. Tycho and Kepler untangled the actual motions of the planets, extracting simple general "laws." Galileo explored and expounded mechanics; he argued and taught and preached to establish realistic science. It was a tremendous advance from earliest watching of the planets and awed marking of eclipses to Galileo's telescope and Kepler's Laws. If ten thousand years seems a long time even for that, remember it is only some four hundred human generations. Are four hundred generations too many for man's understanding to grow from simple superstitions to mathematical certainty? Many of us think this was rapid advance. Yet in the three or four generations of the following century the awakened experimental science made still greater strides.

By 1600 the new growth of science was well under way. Kepler and Galileo were at work. Astronomy was nearly ready to provide a vast frictionless laboratory for mechanics, and experimenting was becoming fashionable. By 1700, planetary laws extracted from observations had already been used to test general laws and theories of mechanics. Newton was developing a new scientific method, in which general theory, devised by shrewd guessing, was made to yield a variety of results by deduction—an ancient method in itself, but now the results were tested by experiment. The deductive method, so dangerous and unscientific when used with wordy arguments alone, was taking its proper place in science, fostering a real alliance between theory and experiment. There were changes in the political, social, and religious structure of Western civilization which gave science greater opportunities. Throughout the century science was blossoming and becoming fashionable; experimenting and realistic argument were all the rage.

Many people helped to build the science of mechanics and to put the backbone of theory into astronomy. Some invented or improved the mathematical tools needed by the physicists—though the latter, like the true blacksmiths, mostly learned to make their own tools to suit their needs. Some carried over the new experimental attitude and the zest for clear thinking into other fields of science; with growing interchange of knowledge, many sciences developed together. Scientists brought fame to their country, and royal favors were bestowed on them from pride instead of superstition. Besides, there was a suspicion that scientists could be useful in commerce and manufacture and war—an early glimpse of their contributions to industry today[5]—so perhaps it paid to favor them!

This was also the time of starting of scientific societies. An Academy of Science was founded in Florence and another in Paris; and the Royal Society was founded in London. These helped science to emerge from the secrecies of the dark ages. They supported some experimenting and encouraged much discussion and interchange of problems and knowledge, but their greatest contribution was publication. No longer were scientific discoveries transmitted by letter to a few friends. They were tested and extended by experiment and argument, then published in print for others to teach and use. With so much keen discussion among able, leisured people, scientific questions were in the air, and the time was ripe for rapid progress.

The names of Copernicus, Tycho Brahe, Kepler,

[4] Dampier, *op.cit.*, p. 108.

[5] Seventeenth-century science is said by some to have laid the foundations of the Industrial Revolution. Others see that later change growing quite independently of scientific attitudes, and only drawing on some factual knowledge.

Galileo, and Newton serve as landmarks, but there were many others who helped to build science in the seventeenth century. Here are short notes on a few of those concerned with physics and astronomy. Others made tremendous advances in biology and medical science (circulation of the blood, mechanism of breathing, embryology . . .).

William Gilbert (1540-1603). Physician. Experimented with magnetism and wrote a very good book on it. Also experimented on static electricity.

Francis Bacon (1561-1626). Brilliant writer who laid down rules for making discoveries by experimenting and induction. These rules are not very practical and did not contribute much to the development of science. He sneered at the work of Gilbert and Galileo and rejected the Copernican theory. However, he did help to spread the idea that nature should be investigated by experiment and not just described by argument.

René Descartes (1596-1650). Philosopher and mathematician. Contributed greatly to Newton's work. Born in France of a rich family, he took life easily yet accomplished much. Did a great deal for philosophy and mathematics, and contributed to anatomy. In physics, he studied optics and motion—going some way toward Newton's Laws—and the nature of matter. He produced an ingenious vortex theory to account for gravity, cohesion, and the motion of the planets. His greatest contribution to physics was his *invention of graph-plotting* with x and y coordinates, leading to the use of algebraic equations for curves, tangents, etc. This prepared the ground for the invention of calculus, which is concerned with drawing tangents and finding areas on such graphs by calculation from their equations instead of by measurement. Our x-y graphs are named "cartesian" after him.

Otto von Guericke (1602-1686). Devised a workable vacuum pump and used it to give demonstrations of atmospheric pressure with the "Magdeburg hemispheres," great hollow cups that teams of horses could not pull apart when the air had been pumped out.

Evangelista Torricelli (1608-1647). Physicist. Made first barometer.

Blaise Pascal (1623-1662). Theologian and scientist. Started the mathematics of probability. Stated laws of pressure for fluids at rest.

Robert Boyle (1626-1691). A great experimenter: vacuum, gas law, chemistry. One of the original members of the Royal Society. Wrote "The Skeptical Chymist."

Christian Huygens (1629-1695). Mathematician and physicist. Developed a wave theory of light. Built a very good clock (probably the first working pendulum clock); and even arranged to correct for slight increase of pendulum period with large amplitude. Studied mechanics, and derived v^2/R for centripetal acceleration earlier than Newton.

Robert Hooke (1632-1702). Began scientific work as Boyle's assistant, but soon rose to be a great scientific experimenter and thinker. His rivalry with Newton obscured his fame and hurt him greatly. But for Newton's overshadowing genius, Hooke would have been classed as one of the great scientists of the age. He bitterly claimed some of Newton's mechanics as his own discovery. Original member of the Royal Society.

Edmund Halley (1656-1742). Astronomer. Friend of Newton. Did much to help the publication of Newton's *Principia*. Important member of the Royal Society.

Science.

There were five important growths in science: (i) the growth in *respectability* of experimental science with increasing freedom of speech; (ii) the growth of *factual knowledge and theory* used in describing it; (iii) the growth of *mathematical tools*; (iv) the invention of *new instruments* for experimenting; and (v) the change in *scientific method and attitudes*.

(i) *Respectability*. We see the growth of science's respectability in Galileo's own life. His father regarded mathematics and science as a poor academic occupation; yet Galileo, rebel though he was, was respected as one of the world's great men in his later years. Newton, Boyle, and Hooke did not have to defend their interest in science; they argued about their discoveries, but not about the spirit of discovery itself. They wrote with little fear of condemnation or ridicule, only with anxiety lest they miss priority or fame. The discussions and publications sponsored by the new societies started scientific knowledge on its way to becoming public and universal—thus the idea of scientific truthfulness began to react on the thinking of mankind.

(ii) *Knowledge*. The growth of actual knowledge was great and varied: e.g. Kepler's Laws, Halley's orbit for his repeating comet, Hooke's Law for springs, Harvey's discovery of the circulation of the blood, Boyle's chemical discoveries and his gas Law.

(iii) *Mathematics*. Coordinate geometry (x-y graphs) was invented, calculus followed, and each helped the other.

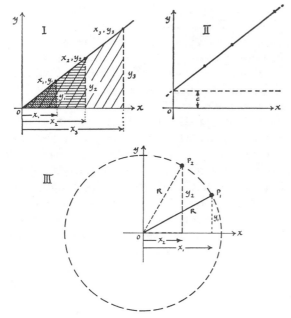

FIG. 20-2. CARTESIAN GRAPHS

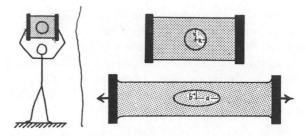

FIG. 20-3. CIRCLE STRAINS INTO ELLIPSE

Graphs link algebra with geometry by converting geometrical forms and operations to compact algebra, and by exhibiting algebra for easy survey.

Graph I shows a straight line through the origin, with several points (x_1, y_1), (x_2, y_2), . . . marked. By similar triangles, the fractions y_1/x_1, y_2/x_2, . . . are all equal—the same for every point on the line. Call the constant value of this fraction k. Then every point on the line represents a pair of values, (e.g. x_1, y_1), which fit the relationship $y/x = $ k or $y = $ k $\cdot x$. This is the algebraic description of the graph, and the line is the geometric picture of the relationship. If y and x are a pair of physical measurements (e.g. s and t^2 for a falling body) the straight line expresses the relationship $y = $ (constant) x, or $y \propto x$, and the slope of the line gives the (constant).

In Graph II every y has a constant bonus, c; so its equation is $y = $ k$x + c$. In this case we can *not* say that $y \propto x$, but only that $\Delta y \propto \Delta x$.

Graph III is a circle, and

for a point P_1,　$x_1{}^2 + y_1{}^2 = R^2$;
for P_2,　　　　$x_2{}^2 + y_2{}^2 = R^2$,

so the equation of this circle is:

$$x^2 + y^2 = R^2.$$

Or we can write this: $\dfrac{x^2}{R^2} + \dfrac{y^2}{R^2} = 1$.

An ellipse can be made by distorting a circle with uniform strain. Draw a circle on a sheet of rubber and stretch it. Radii R and R become semi-axes a, b.

Then, at a guess, a circle with

equation $\dfrac{x^2}{R^2} + \dfrac{y^2}{R^2} = 1$　and AREA $= \pi R^2 = \pi R \cdot R$

becomes an ellipse with

equation . . ? . $= 1$　and AREA $= $?

Thus, coordinate geometry could reduce orbit ellipses to equations that are easier to handle.

Science had developed two serious mathematical needs, both filled by calculus: to *calculate* tangent-slopes of curved graphs, and to *calculate* areas inside curves, *by algebra*. *Differential calculus* does the first, and the reverse process, *integration*, does the second. A tangent-slope gives a rate-of-change. Calculus is just an algebraic process to find a rate-of-change *at an instant*. It enables us to calculate ACCELERATIONS from an equation that specifies VELOCITIES, or VELOCITIES from an equation relating DISTANCE and TIME. (For example: given $s = 16t^2$, calculus tells us that $v = 32t$; and it then tells us that $a = 32$, a constant value.) Integration—again just a refined algebra-logic-machine—adds up an infinite number of infinitely small contributions: tiny patches to find an area (e.g. for Kepler's Law II), or tiny attractions to find a total gravitational pull.

You used graphs and calculus for your early investigation of a wheel rolling downhill:

Stage 1. EXPERIMENT → EMPIRICAL GRAPH. You plotted s against t^2, each point coming from an experimental timing. *The points are the facts.* A line sketched to follow them is a summary of facts—in the grammar of graphs, an "indicative" graph.

Stage 2. THINKING → THEORY. Guess at constant acceleration, as a possible simple rule for nature. Use calculus to predict the necessary relation between s and t. Integration adds all the little distances travelled with growing speed, and shows that *if* acceleration is constant, *then* s must vary as t^2.

Stage 3. TEST. Draw a straight line through origin to represent the relationship $s \propto t^2$ from theory.

IF your points lie close to this line, THEN *the wheel's motion in the experiment is close to constant acceleration.* That straight line on your graph is an "interrogative" line: drawing it is asking a question, "do the facts fit constant acceleration?" You would be wise to call such a graph line a "question-asking line." Drawing the "best straight line" on your graph of an experiment exhibits your *hypothesis*—a temporary rule, guessed at and to be trusted for the moment (e.g. "suppose $\Delta v/\Delta t$ is constant . . .").

(iv) *Instruments.* A new instrument, as well as a new mathematical tool, can promote great developments of science. The 17th century was a century of great inventions of instruments: the telescope, the microscope, the vacuum pump, the barometer, the pendulum-controlled clock, and the first crude thermometers were the tools that promoted a new range of experimental science.

(v) *Attitude and Method.* From the Greeks to Galileo, science was being built by collectors, accurate observers, makers of schemes, authoritarian philosophers.[6] The collectors gathered a lot of knowledge which by itself would have been too diverse to be called science. The scheme-makers organized this knowledge and extracted rules that were good working prescriptions, able to summarize the facts and often to make predictions. Rules and knowledge, together with techniques for gaining more knowledge, made the beginning of the new science.

Meanwhile the thinkers were busy devising explanations—statements that would make knowledge fit together better and become easier to "understand" or easier to accept. Many explanations or reasons were drawn from their own thinking with only remote experimental background—e.g., for epicycles, "circles are perfect"; for the barometer, "invisible threads hold the mercury up." Some explanations seem to be little more than a statement that nature "is like that," put with authority—e.g., for falling bodies, "the lowest place on the ground is *natural.*" Man needed such reassurance that the external world of nature has a simple organization in it; otherwise his fears of a profuse unknown would have driven him to more superstition and

[6] We may trace some survivals from childhood in these activities: a child's delight in collecting may turn to scientific data-collecting instead of adult stamp-collecting or even money-hoarding; the grim determination of many adolescents not to be beaten by a hard problem may turn to a Tycho's drive for accuracy, instead of to some more cruel behavior; the insecure child's craving for a definite framework of rules may turn to hunting for simple laws instead of to some more neurotic form of worry.

even madness. As general working rules emerged—e.g., the epicycle scheme, Hooke's Law, Kepler's Laws—the sense of security and comfort increased, and the early belief that nature is definite and reasonable gained ground as a basic belief in science. The Greeks *deduced* their explanations and schemes for nature from a few general ideas which they just assumed—e.g., from "circles are perfect" they deduced epicycloids. In the course of the 17th century that kind of deductive reasoning fell into disfavor; it was really philosophical speculation flourishing with authority, rather than science. By the middle of that century experiment was regarded as the real source and test of science. Men were occupied with extracting rules or laws by inductive reasoning from experiment. In doing this, they too were making assumptions: that nature is simple, and that nature is uniform—that is, that in the same conditions the same behavior will occur again and again. They still assumed that there are *causes* for things, but the meaning of causality remained as difficult a problem as ever.

Though this inductive method was an honest one leading to good rules, it lacked the general tying-together and mental satisfaction that a grand theory can give. Newton, with greater insight, looked at experiment then jumped to theory and worked back deductively from theory, predicting results that could be tested. This brought theory back into science as a framework of thought, but in a more respectable and responsible form. Theory was again considered valuable—e.g., the theory of universal gravitation—but *as a servant to science rather than as master.*

Later still, say in the last century, theory was subjected more and more to the test of productiveness. Scientists asked, "Can this theory make (further) predictions?" If not, it was shelved or modified. That now seems too harsh a treatment for theory. Its use may lie not only in its ability to make predictions but also in our enjoyment of the scheme of thinking that it offers us.

Descartes' New Philosophy

While the viewpoint of working scientists was thus changing with the temper of the times, René Descartes in France advanced a new philosophy that had a lasting effect on scientific thinking. And with it he announced a new model of the universe that remained popular for a century. Seeing faults in classical philosophy, Descartes turned his consideration of the world inward to his own thoughts and feelings and set himself to doubt every stage of his knowledge. From this examination he evolved a *dualism*, a picture of two sharply-divided worlds existing together, *each as real as the other*: a world of matter, with size and shape and motion, and a world of soul and mind. Just as two clocks side by side can keep the same time, so the two worlds, although entirely separate, keep in tune, because "God made them so."

In this scheme, matter is entirely dead, without spirit, and it can only exchange motion with other matter by contact. The motion of matter must have been started originally by God. Thus, to Descartes, God was not a presiding power who controlled the world for man's life in it; but God was the First Cause, who started the Universe with motion, laid down rules for its running, and then left it to run. Thereafter, motion can only be carried from one piece of matter to another by some matter. Therefore, the open spaces in the solar system cannot be empty. They must be full of an invisible material "aether" that carries the motion. Since a moving region of aether cannot continue out indefinitely, it must be arranged in a closed circuit, a whirlpool or *vortex*. All space in fact is full of vortices of aether, large and small all geared together, conveying the motions of visible bodies. The planets are carried around their orbits by a huge vortex belonging to the Sun. The Earth, carried with the other planets in that vortex, has a smaller vortex of its own to draw objects inward. Thus, inward fall under gravity corresponds to the sweeping of floating straws towards the center of a whirlpool in water. On a smaller scale the picture accounted for cohesive forces that hold small pieces of matter together. This scheme of whirlpools within whirlpools, all invisible, sounds fantastic today; yet it proved very popular at the time, because it explained the whole system of the universe by a vast machine started by God but then kept running by constant mechanical rules. In fact, Descartes' picture of a "full" universe with no vacuum, running as a machine, became a serious rival when Newton published his gravitational theory. Newton favored a vacuum and gave no ultimate cause for gravitation. Descartes' theory offered more explanation, but it rejoiced in unsupported speculation—the vortices were undetectable except by the motions they were to explain. Newton attacked them mathematically, showing they could not fit with Kepler's Law III, and he attacked them on principle when he claimed, "I will not feign hypotheses."

Out of his systematic doubting, Descartes seems to have emerged with a certainty of God. God started the universe and provided laws to govern it. Therefore, the laws of nature must be completely right: God would neither make a careless mistake nor accept a rough average. This view of the laws of nature influenced the next generation of scientists— Newton and his contemporaries felt they were looking for great laws established by God's command and waiting to be found.

If you think this a strange digression in a discussion of hardheaded physics, reflect that the same problems remain today at the borders of science and philosophy: What is the nature of space (that carries electromagnetic and gravitational fields, and obeys Relativity geometry)?; What do the laws of nature mean?; How did the world begin?; How old is time?; Will time go on?

Francis Bacon

While Descartes tried to explain the universe with a grand *deductive* theory operated by mathematics, Francis Bacon in England advocated a grand *inductive* treatment of systematic experimenting. He sought universal knowledge by great organized schemes of research, to produce stores of data from which scientific knowledge could be extracted. He claimed that science could not advance by pure deduction and argument, nor could it advance rapidly by haphazard data collecting—children's play. Rather, scientists should plan their experimenting carefully and treat it by a formal system of inductive reasoning and testing.

Thus, Bacon saw clearly the difference between "good experimenting" and "just playing around with apparatus." He set forth an ideal scheme for science: collect information, extract rules, frame hypothesis, deduce consequences, test deductions.

However, if you watch scientists at work you will see there is no one scientific method. Physical science does not develop as a simple rigid chess-game of alternating moves; there is far more variable and complex interplay of pieces, players, and the board itself. Nor is the progress just a series of forward leaps. A first round of thinking and experimenting may even lead back to the starting point—"this is where we came in"—but, as with seeing a movie over again, we have a richer knowledge with which to pursue the second round. As J. R. Zacharias puts it, "science bootstraps itself."

Bacon wrote with great eloquence, advocating a vast organization of professional experimenters and reasoners. His grandiose scheme was too artificial for successful science, and it aimed more at practical values than ultimate understanding of nature. (It had some of the misplaced zeal that we might expect today from a non-scientist research-director placed in charge of a large commercial research laboratory.) His proposals were only schemes on paper, yet they had great influence on the starting of the Royal Society and the experimental work of its members, in particular Boyle. By mid-century, "art was giving way to science under the pressure of Bacon's influence, . . ."[7] Nowadays, two centuries later, we see that in a deeper sense science *is* an art.

The Growth of Theory: A Necessity for Modern Science

1600 to 1700 was an exciting century for astronomy—as for all the life of the intellect. At its begin-

ning, there was a growing stock of facts and rules, pressing for explanation. Speculations on general causes were in the air: the time was ripe for a comprehensive view. By the end of the century, knowledge and interest had grown and spread; but, above all else in importance, Newton had built and published a great theoretical scheme that pulled astronomy into a single "explanation" and projected it into a future of rich promise.

If you wish to understand modern physical science, you need to know what theory does; you need to have a feeling for "good theory." You can hardly learn that from sermons *about* theory. Instead, study a good example of it. The next four chapters describe and discuss Newton's great theory of gravitation.

[7] Dorothy Stimson, *Scientists and Amateurs, A History of the Royal Society* (Abelard-Schuman Ltd., New York, 1948), p. 36.

CHAPTER 10 · VECTORS AND GEOMETRICAL ADDITION

"What hopes and fears does the scientific method imply for mankind? I do not think that this is the right way to put that question. Whatever this tool in the hand of man will produce depends entirely on the nature of the goals alive in this mankind. Once these goals exist, the scientific method furnishes means to realize them. Yet it cannot furnish the very goals. The scientific method itself would not have led anywhere, it would not even have been born without a passionate striving for clear understanding."

—A. EINSTEIN, *Out of My Later Years*

A Personal Experiment

Galileo tried to separate the up and down (vertical) motion of a projectile from its horizontal motion. Experiment vouches for this treatment by showing that these two motions are independent. Try this yourself. Throw one stone out horizontally and at the same moment release another to fall

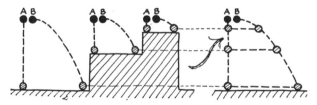

FIG. 2-4. EXPERIMENTAL COMPARISON OF MOTIONS: inferring the general result that falling stone and projected stone keep level all the way.

vertically. They both hit the floor at the same instant. Stone B moving in a curve has to fall the same *vertical* distance to reach the floor as stone A falling vertically. They take the same time. Do A and B keep abreast at intermediate stages of their fall? You need not place special observers to sight them at various levels. Instead, you can move the floor up to catch them earlier and repeat the experiment. Or, more easily, you can move the starting point down nearer to the floor. If A and B arrive at the same instant whatever height they start from, you can say fairly that they keep abreast all the way down. Notice how a series of experiments can be used to replace a difficult complex of simultaneous observations. In trusting our inference from such a set of experiments, we assume the "Uniformity of Nature."

Here are "rules" for the motion of ideal projectiles, without air-friction:

(I) *the motion is independent of the size or mass of the object,*
(II) *the vertical and horizontal motions are independent of each other,*
(III) *the vertical motion has a constant downward acceleration, the same as that of any falling body,*
(IV) *the horizontal motion continues unchanged.*

A projectile does not really have separate horizontal and vertical motions. As it moves along its curved path, its motion at any instant is directed along the tangent. While it rises from A to B to C

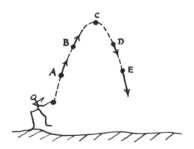

FIG. 2-14. PROJECTILE MOTION

it moves slower and slower, then falling from C to D to E it moves faster and faster, the speed changing as the vertical part of the motion is changed by "gravity."

The splitting up of the actual motion along the path into horizontal and vertical motions, called *components*, is an artificial trick. The process of compounding separate motions into a single motion which we call the *resultant*, is important in navigation where motions of ship and ocean currents, or plane and wind, are to be combined. In the next section we shall study such adding of motions.

Geometrical Addition

No one watching the curved flight of a stone flung in the air would automatically separate it into vertical motion and unchanging horizontal motion; yet as scientists we are encouraged to make this separation—analysis—when we discover that the two motions are of different types and are independent of each other. Attempting such analysis at once raises the questions: (i) How is a single slanting motion split up into two ingredients or "components"? (ii) How are two separate motions to be compounded together into one single motion? We can guess the answer to the second question, and use it to answer the first. If we try to add two or more *motions*, we have to keep track of simultaneous movements in different directions. Instead, let us allow the motions to proceed for some specified time, say one hour, and then deal with the *distances travelled* in that time. Then the problem of adding motions becomes a simple one of adding travelled distances, or journeys or trips.[1] Is the addition rule the same as in arithmetic, as in adding 2 and 3 to make 5?

Experiment soon shows us this will not work unless the separate journeys to be added are straight ahead in the same direction. Then we see 4 ft due North and 3 ft due North do make a total trip of 7 ft due North as in Fig. 2-15. (And, therefore a

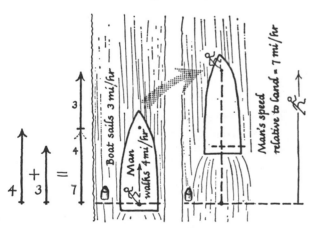

FIG. 2-15. ADDING MOTION IN SAME DIRECTION

speed of 4 ft/sec and a speed of 3 ft/sec both due North do make a total speed of 7 ft/sec due North. And 4 miles/hr plus 3 miles/hr both in the same direction do make a total speed of 7 miles/hr.)

However, if the directions are different simple arithmetic does not work. A trip of 4 ft due East

added to 3 ft due North does not make a trip of 7 ft. Nor does a speed 4 miles/hr due East plus a speed 3 miles/hr due North make a speed of 7 miles/hr in any direction. To fit the facts of the world, we have to use another kind of addition, which we call *geometrical addition*. Common sense—in this case simple knowledge accumulated in crawling, walking, driving, sailing, etc.—suggests how geometrical adding should be done. Suppose you wish to add trips of 4 ft to the East and 3 ft Northward, to find the *single trip that would carry you from the starting point to the destination*. Though it seems childish, try this for yourself. Stand facing North with your feet together. Then try to make both these trips, i.e., step four paces to the right and three paces forward at the same time. You could try this by doing one trip with each foot; sideways with your right foot and forward with your left foot, simultaneously; but the result

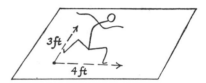

FIG. 2-16. TRYING TO ADD TWO MOTIONS IN DIFFERENT DIRECTIONS

is uncomfortable (Fig. 2-16). Instead you had better take one trip first, then the other, thus: move 4 paces to the right *then* 3 paces forward (Fig. 2-17). Or you can reverse the order and ar-

FIG. 2-17. ADDING MOTIONS

rive at the same destination. If you could somehow make the two trips simultaneously you should reach the same end-point. In fact this can be done if you have a rug which can be drawn across the floor by an electric motor. Then have the motor drag the rug with you on it (or a toy, as in Fig. 2-18) 4 paces to the right while you move 3 paces forward at the same time. On the rug—relative to the rug—you only move 3 paces forward. From a bird's eye view you make both journeys simultaneously and reach the same destination as if you made first one journey then the other. What single trip could replace these two, whether they are taken simultaneously or separately, and get you to the same destination? The simple single

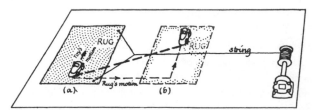

FIG. 2-18. ADDING MOTIONS. The toy crawls along the rug while an electric motor pulls the rug across the floor. The toy has a diagonal motion over the floor.

trip is along the straight line from starting point to finish. This is called the *resultant* of the two trips. If the trips are drawn to scale on paper, as in Fig. 2-19, then the single trip which would replace

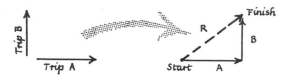

FIG. 2-19. ADDING PERPENDICULAR TRIPS

them (if they are taken separately) is trip **R**. If the trips are not at right angles, a similar scale drawing will work, as in Fig. 2-20. If the trips are taken

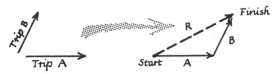

FIG. 2-20. ADDING TRIPS

simultaneously—as when a plane flies in a wind—we can still pretend to take first one then the other, and arrive at the resultant **R**, as in Fig. 2-21.

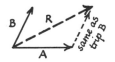

FIG. 2-21. ADDING TRIPS

We find the resultant by taking first one trip then the other, as in Fig. 2-22a or Fig. 2-22b. Combining these figures in Fig. 2-22c, we see that the

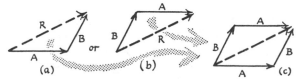

FIG. 2-22. ADDING TRIPS

resultant is given by the diagonal of the parallelogram whose sides are the original trips.

This system is obviously right for adding trips: we are assured by common sense, based on ex-

perience ranging from nursery exploration to complex navigation.

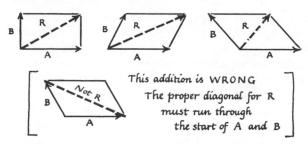

FIG. 2-23. EXAMPLES OF PARALLELOGRAM ADDITION

The system can be reversed, and the trip **R** split into components **A** and **B**. They are one possible pair that would combine to make **R**. There are an infinite number of such pairs, each adding to the same **R**.

Velocity and Speed

The *direction* of a motion is just as important as its size. We now need a name for the idea of a *definite speed* associated with a *definite direction*. We call this *velocity*.[2] Velocity then has two qualities: size (= speed) and direction. Do velocities add by the geometrical system? Or, as a scientist would say, are velocities "vectors"?

Vectors: Definition

Vectors are those things which are added by the geometrical system. They are called "vectors," because we can draw a line to represent them, showing both their size (to some scale) and their direction.

RULE FOR ADDING TWO VECTORS

The following rule describes geometrical addition. Our definition of vectors makes it automatically true for vectors.

Geometrical addition: *To add two vectors, choose a suitable scale, and draw them to scale starting from the same point. Complete the parallelogram. Then, on the same scale, their resultant is represented by the diagonal from the starting-point to the opposite corner.*

In this, the *resultant* of a set of vectors is defined

---

[2] In ordinary language, speed and velocity mean the same thing: how fast an object is moving. In physics, it is useful to reserve the name velocity for speed-in-a-particular-direction, which is a vector. From now on, we shall use speed to mean just rate of covering distance along some path whether straight or crooked—a worm's measure of progress. A speed is specified by a number with a unit, such as 15 miles/hour. A velocity needs a number with a unit *and* a direction to specify it, e.g., 15 miles/hour Northward.

as that single vector which can replace, or has the same physical effect as, the original vectors taken together.

FIG. 2-27. "TAIL-TO-HEAD" ADDITION. Adding two vectors by parallelogram method is equivalent to "tail-to-head" addition.

Starting with parallelogram addition, we can omit part of the drawing and still obtain R.

We can economize still further and draw only a triangle, and we are back to our first discussion of trips, where we added them by taking first one trip and then the other. This leads to an easy *rule for adding vectors*:

DRAW ONE OF THEM FIRST.

THEN DRAW THE SECOND, STARTING IT WHERE THE FIRST ONE ENDED—that is, draw them one after the other, "tail-to-head."

THEN DRAW THE LINE JOINING START TO FINISH, AND THAT REPRESENTS THE RESULTANT, R.

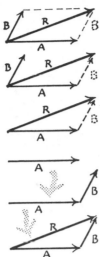

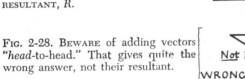

FIG. 2-28. BEWARE of adding vectors "*head*-to-head." That gives quite the wrong answer, not their resultant.

Which things are vectors? That is, which things in science do add geometrically by the parallelogram construction? Trips, or to give them a more official name "directed distances" or "displacements," are vectors. If trips are vectors, we need only divide by the time taken to travel them to see that velocities are vectors too. If we use as vectors the length travelled in unit time, then these vectors, which add geometrically as trips, themselves represent velocities. As an extension of this, we see that accelerations are vectors too.[3] We shall find other vectors, other things that can be measured with instruments and which obey geometrical addition. At the moment an important question arises: are forces vectors, i.e., do they obey geometrical addition? This cannot be answered by thinking about it.[4] It is not obvious. It needs experimental investigation. Experiments show that forces *are* vectors. They do add by geometrical addition to give a resultant—a single force that can replace them.

---

[3] *Trips* are vectors. *Velocities* are *trips per hour*, say. Therefore velocities are vectors. Therefore *changes of velocity* (which are themselves each a velocity gained or lost) are vectors. *Accelerations are changes of velocity per hour*, say. Therefore accelerations are vectors.

[4] Unless we are prepared to *define* forces as things which add geometrically and then take the consequences of our definition in later developments.

Scalars and "Super-vectors"

Things which are not vectors but have only size, without any direction attached, are called *scalars*; for example, volume, speed, temperature. There are other things which are neither vectors nor scalars; some vague things such as kindness, and definite ones, some of them "super-vectors" called tensors. The stresses in a strained solid provide an example of tensors: pressure perpendicular to any sample face and shearing forces along it. More complicated examples appear in the mathematical theory of Relativity. For example, we shall treat momentum, mv, as a vector with three components, mv_x, mv_y, mv_z; and we shall treat kinetic energy as a scalar. Einstein, taking an overall view of space-time, would lump momentum and kinetic energy into a "four-vector" with four components, three for momentum, one for kinetic energy.

Projectiles and Parabolas

We can analyze the shape of a projectile's path with the help of geometry.

Geometrical analysis: Suppose a stone is thrown out horizontally. Then its horizontal motion carries it the same distance horizontally every second, while it accelerates vertically. It falls 16 feet vertically in the first second from its start, 64 feet in the first two seconds, 144 in the first three seconds, etc. Make a scale map of its positions at several instants of time. Choose total times from the start which run in the proportions 1:2:3:4. . . . In these times it travels steadily sideways, covering distances in the same proportions 1:2:3:4 . . . . ; but it falls vertical distances proportional to the squares of these numbers, to 1, 4, 9, 16 . . . because

$$\text{VERTICAL FALL} = \tfrac{1}{2}g\,(\text{TIME})^2$$

and $(\text{TIME})^2$ has values in proportions 1:4:9. . . . Map its position at these equally spaced instants of time by drawing vertical lines evenly spaced, say at intervals of 2 inches across the map; and horizontal lines 1 inch down from starting level, 4″ down, 9″ down, and so on to show vertical falls. Then the predicted path is marked by the crossing of these lines, as shown in Fig. 2-36. This can be tested by throwing marbles or coins in front of a wall on which such lines have been ruled.

If the stone is not thrown out horizontally but is flung up along a slanting direction, the story is similar. The initial slanting motion given by the thrower remains unchanged during flight while an increasing vertical falling motion, due to gravity, is added to it.

110

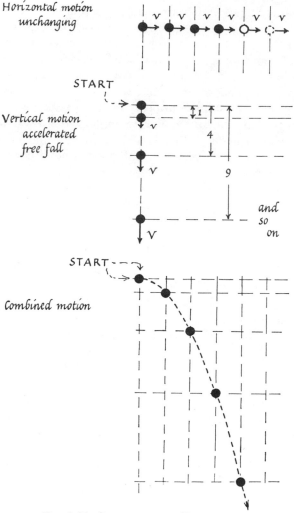

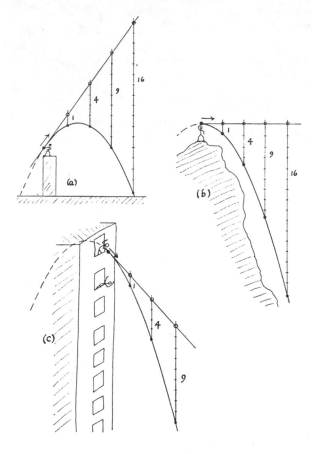

then falls when the accelerated rate of free fall has beaten the steady speed of rise due to the initial motion. (See Fig. 2-41. Note that this path is a parabola.)

FIG. 2-36. COMBINING THE HORIZONTAL
AND VERTICAL MOTIONS OF A PROJECTILE.

FIG. 2-41. FREE FALL OF A PROJECTILE.
However it is started, a projectile falls with the same "free fall" from its original starting-line as an object released from rest. The accelerated motion of fall is independent of both the vertical and horizontal components of the initial motion.

But for air-friction, any projectile drops freely from its starting line, from the very beginning. It falls 1, 4, 9, 16, . . . feet in 1, 2, 3, 4, . . . quarter-seconds from start. If the starting line slants upwards, the projectile's actual path rises at first and

Notice how our discussion has torn the problem of projectile motion to pieces, leaving it easier to deal with, ready for further studies by experts in ballistics. We have not so much set forth new information as made existing knowledge easier to use.

". . . all science as it grows towards perfection becomes mathematical in its ideas."
—A. N. Whitehead (1911)

The Problem of Orbital Motion

Why do the planets move in Kepler orbits? Why do they keep moving, and why are their orbits ellipses? These questions naturally followed Kepler's discoveries, in the long tradition of asking WHY, dating from the Greek philosophers. Astronomers had measured and recorded WHAT the planets do. Copernicus and Kepler had shown HOW the planets' motions can be expressed in a simple scheme; but they had only the traditional answers to WHY. Copernicus thought of spheres rolling around, though with simpler motion; Kepler imagined a spoke of influence from the Sun carrying each planet and pushing it *along its orbit*; and he talked mystically of magnetism shaping the orbits. His spokes were real in one sense; they were needed to express his

---

PROBLEM ON SUBTRACTING VECTORS (Problem 1.)

 We shall need to subtract vectors in studying planetary motion. This problem is to give you practice.

I. <u>Ordinary</u> (arithmetical) <u>subtraction</u>. Suppose we want to subtract 2 from 5.

 This can be regarded in several ways:

 (a) We can say, 2 subtracted from 5 makes . . . . . . _____

  or, the same thing in other words, 5 - 2 is . . . _____

 (b) Or we can change the sign of the 2 to -2 and ask an "addition" question: 5 + (-2) makes ? . . . . _____

 (c) Or we can put this more childishly and ask: What must we add to 2 to make 5 ? . . . . _____

This last form gives the key to subtracting vectors (or finding the <u>difference</u> between two vectors, or finding the <u>change of vector</u> from one vector to another.)

II. <u>Vectors.</u> Suppose we have an "old vector" and a "new vector" and want to find the <u>change</u> (or <u>gain</u>, or <u>difference</u>). We ask, "What vector must be added to the old one to make the new one?" (This is like form I(c) above; but it requires <u>geometrical</u> addition.)

 (a) If the vectors both point due East as below, old vector 2 and new vector 5, what is the change or difference? "What must be added to the vector 2 to make the vector 5 ?" <u>Show it by drawing on the sketch below.</u>

                     FIG. 21-1.

New vector 5

Old vector 2

 (b) If the vectors (old vector A and new vector B) have different directions, as in the various cases shown below, what is the change or difference? "What must be added to the vector A to make vector B?" <u>Show it for each case by drawing an arrow, on this sheet.</u> In each case, you want B - A.

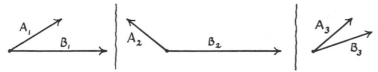

 (c) If the vectors do not sprout from a common starting point, you must first transfer one or both till they do. Then find difference, B - A, in each case below, <u>again using an arrow to show your answer.</u>

                      FIG. 21-2.

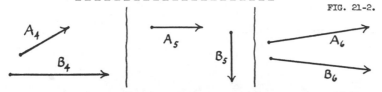

Law II; and they are still with us, as geometrical lines that sweep out areas. As solid arms to propel the planets Kepler's spokes soon seemed unnecessary: Galileo's new teaching put things in a different light. A moving thing, said Galileo, will continue if it is left alone, and he gave a clever thought-experiment to justify his view. A generation later, Newton expressed this view as a working rule: Law I, "Every body will remain at rest or continue to move with constant speed in a straight line unless acted on by a resultant force." Then Newton crystallized the vague idea of motion into definite momentum, to be calculated by multiplying mass (which he tried to define) by velocity, and said: Law II, "When there is a resultant force, there are changes of momentum in the *direction of the force*. RATE OF CHANGE OF MOMENTUM varies directly as the RESULTANT FORCE." This was equivalent to "MASS · ACCELERATION varies as RESULTANT FORCE."

From Galileo to Newton these new views of motion—groped for by philosophers far earlier, half stated by Leonardo long before Galileo and by Descartes after him—were getting ready to play a profound part in astronomy. The members of the newly-formed Royal Society, which soon welcomed Newton as a younger Fellow, discussed Kepler's laws eagerly, asking quite a different kind of WHY question. They no longer worried about an agent to push the planets *along*. Galileo told them no push—and therefore no pusher—is needed for that; the planet will continue to move of its own accord if left alone, like a block of ice on a frozen pond or a bullet in space. Scientists had dissolved away Kepler's spoke. What they sought instead was an *inward* force, along the spoke perhaps, to pull the planet into a curved orbit instead of a straight line. Such a force pulling sideways, "across the motion" of the planet, would give it momentum in a new direction. What kind of forces would do this? This new question was in the air. Hooke, Huygens, and Newton all attacked it. Taking the planetary orbits as roughly circular, they argued back from Kepler's Law III and suggested that there is an inward attraction, pulling planets to the Sun, a pull that decreases with increasing distance according to an "inverse square law"—a scheme we shall discuss in the next chapter. But would such a force, half suspected and quite unexplained, produce elliptical orbits fitting with Kepler's Laws I and II? This was too difficult for all but Newton. It required clear formulation of laws of motion, and then clever mathematics. Newton not only solved it but ex-

tended his solution into a magnificent framework of good theory. Before you study his work, you should extend your discussion of force and motion to this new case of "sideways forces" that pull a moving body's path into a curve. You have already met this with projectiles, where gravity adds vertical momentum to horizontal motion, making a curved path.

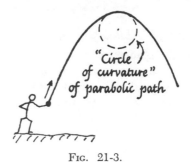

FIG. 21-3.

But acceleration in that motion with its increasing speed seems easier to understand—dare we say it "seems more natural"?—than in steady motion around a circular orbit, with unchanging speed.

Acceleration of Body Moving Around a Circle

Suppose we have a planet moving around a circle (or a stone on a string, or an airplane, or an atom). Does it have any acceleration? If not, we hardly expect to find any resultant force acting on it; and in that case why does it not continue straight ahead? Does it have any acceleration? Certainly not any acceleration *along* its path—we have chosen a case of constant speed, with no speeding-up along the path. Is there any acceleration *across* the path, perpendicular to it? Try drawing vectors to look for changes of (vector) velocity. The moving body P

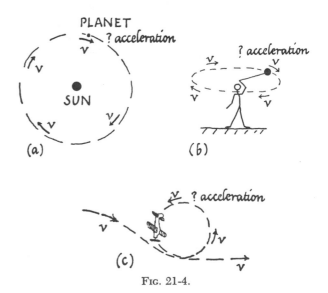

FIG. 21-4.

travels around a circle of radius R with fixed *speed* v.[1] Then v is also the size of P's *velocity*, but that velocity, as a vector, has direction as well as size and the direction changes from instant to instant. When the moving body, P, is at A it has velocity v in the direction shown, along the tangent. If it moves with unchanging speed, the *size* of v is the same at B as at A, but its *direction* is different; the two vectors are not identical. There *is* a change of velocity between A and B. (And therefore an acceleration, and therefore . . . on to planetary astronomy with a wonderful tale, if this is so.) Calculate the change of velocity and divide it by the

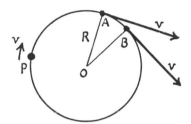

FIG. 21-5. VELOCITY VECTORS

time taken, to find the "acceleration." This involves subtracting vectors to find a change as in the preliminary problem at the beginning of this chapter.

Derivation of Expression $a = v^2/R$

While P moves from A to B it changes its velocity from (v along AT) to (v along BT'). Make a vector diagram to find the change of velocity. Transfer the two velocity vectors to a common starting point, X, and draw XY to represent VELOCITY v at A and XZ to represent VELOCITY v at B. Then XY is the "old velocity" and XZ the "new velocity." What is the change? What velocity must be added to the old to get the new? The change is shown by YZ, the vector marked Δv in the sketch. Then

(old v) $+ \Delta v$ makes (new v), *by vector addition*

To see the direction of Δv, redraw the original picture, and slide the v's along their tangent lines till they both sprout out from the common point C. Then we can treat C like X and draw old v and new v from it and mark Δv. Look at the direction of Δv. It is parallel to CO, from C to the circle's center. If we took B very close to A, the Δv would have to be along the radius from the region AB to the center. Δv is an inward velocity vector, towards the center of the circle.

---

[1] Remember we use "speed" for rate of travel along any path, straight or curved—the drunkard's speedometer-reading. Speed is a scalar. Velocity is a vector, with direction as well as size.

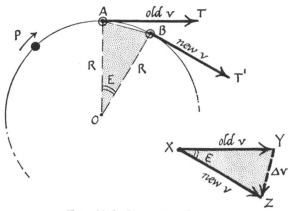

FIG. 21-6. VELOCITY CHANGE
Since velocities are along tangents, which are perpendicular to radii, the fact-picture triangle OAB and the vector-diagram triangle XYZ are similar.

If there are velocity changes, there is acceleration.[2] Calculate the acceleration by dividing velocity change, Δv, by the time it takes, Δt. Time Δt is the time taken by P to move from A to B with speed v along the orbit. In fact SPEED v is $\dfrac{\text{arc } \widehat{AB}}{\Delta t}$. To calculate $\Delta v/\Delta t$ in terms of v and R, etc., we need some geometry discovered by Newton's contemporaries. Here it is. Join A and B by the chord $\overline{AB}$. As often in solving geometrical problems, the trick is to add one construction line, here the chord $\overline{AB}$. Now look for similar triangles between the fact-picture and the vector diagram of velocities. Radii OA, OB in the fact-picture make a small angle E. The velocity vectors are along tangents, perpendicular to the

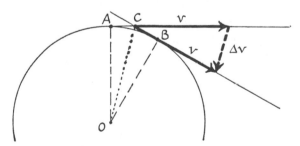

FIG. 21-7. DIRECTION OF VELOCITY CHANGE
Here the vectors for old v and new v have been slid along their lines in the fact picture so that they both start from C. The change of velocity, Δv, is parallel to CO. The change of velocity during travel from A to B is inward towards the center.

---

[2] From now on we shall stop showing **v** and **Δv** as vectors by boldface type because we are going to calculate the *size* of the acceleration, using the *speed* v, which is the *size* of the velocity, and Δv, the *size* of the change of velocity. Remember, however, that the acceleration has the direction of the vector **Δv**.

114

radii, so the velocity vectors (old v) and (new v) make the *same* small angle E.[3] Then in the fact-picture we have triangle OAB with two equal sides, R and R, enclosing angle E; and in the vector-diagram we have triangle XYZ with two equal sides v and v, enclosing the same angle E. Therefore the two triangles OAB, XYZ are similar.

Then $\dfrac{\text{"short side," } \Delta v}{\text{"equal side," } v}$ must $= \dfrac{\text{"short side," } \overline{AB}}{\text{"equal side," } R}$

in vector triangle — in fact triangle

$$\therefore \frac{\Delta v}{v} = \frac{\overline{AB}}{R} \qquad \therefore \Delta v = \frac{v \cdot \overline{AB}}{R}$$

Now we can calculate the "acceleration."

$$\text{ACCELERATION} = \frac{\Delta v}{\Delta t} = \frac{v \cdot \overline{AB}}{R} \bigg/ \Delta t = \frac{v}{R} \cdot \frac{\overline{AB}}{\Delta t}$$

To go further we need to know what $\overline{AB}/\Delta t$ is. What *is* $\overline{AB}/\Delta t$? What is the fraction [(CHORD $\overline{AB}$) divided by (TIME OF TRAVEL from A to B)]? We know what (arc $\overset{\frown}{AB}$)/Δt is. That is distance/time, around the orbit from A to B, so it is the SPEED v. But, for a very short arc, with B very close to A, the curved arc $\overset{\frown}{AB}$ is very nearly the same as the chord straight across, $\overline{AB}$. Look at the series shown in Fig. 21-9. As we move A and B closer and closer together the arc $\overset{\frown}{AB}$ and the chord $\overline{AB}$ get smaller but they also show *much* less difference from each other. Like mathematicians inventing calculus, we crawl towards the "limit" when B coincides with A. We never get to that limit, but we can crawl as close as we like and make the difference between arc and chord as trivial as we like. We do not merely make the *difference*, $\overset{\frown}{AB} - \overline{AB}$, trivially small: we make the *fraction* (difference/chord), or $(\overset{\frown}{AB} - \overline{AB})/\overline{AB}$, trivial. This makes the ratio $\dfrac{\overset{\frown}{AB}}{\overline{AB}}$ very nearly 1. So,

with a big separation between A and B, we can say *arc is somewhat greater than chord*; with small separation we can say *arc = chord approximately*; with still smaller separation, *arc = chord very nearly*; and we can get as near as we like to the limit *arc = chord*. Mathematicians prefer to describe this limit thus: Limit $\left(\dfrac{\text{arc}}{\text{chord}}\right) = 1$. Now we want the acceleration *at an instant of time*, when B and A do practically coincide—we do not want a vague average over a long separation. We want the *limit* when B coincides with A. So we say: in the limit, arc = chord, $\overset{\frown}{AB} = \overline{AB}$.

Then $\dfrac{\text{arc}}{\Delta t} = \dfrac{\text{chord}}{\Delta t}$

or $\dfrac{\overset{\frown}{AB}}{\Delta t} = \dfrac{\overline{AB}}{\Delta t}$, in the limit.

Then acceleration $= \dfrac{\Delta v}{\Delta t}$

$= \dfrac{v}{R} \cdot \dfrac{\overline{AB}}{\Delta t} =$, in the limit, $\dfrac{v}{R} \cdot \dfrac{\overset{\frown}{AB}}{\Delta t}$

$= \dfrac{v}{R} \cdot (v)$ since $\dfrac{\overset{\frown}{AB}}{\Delta t}$ is v.

Then the acceleration $\dfrac{\Delta v}{\Delta t} = \dfrac{v}{R} \cdot (v)$

$= \dfrac{v^2}{R}$ or $\dfrac{(\text{SPEED AROUND ORBIT})^2}{\text{RADIUS OF ORBIT}}$

This relation, ACCELERATION $= v^2/R$, is of great importance. We shall use it in planetary theory, in studying electron streams, in making a mass spectrograph for atoms, in designing a cyclotron—in fact wherever we meet motion around orbits. It is so important that you should retrace its derivation for yourself and make sure it is sensible. Once you understand the derivation you will see that you can reduce it to a short explanation + two sketches + a few lines of algebra.

Two Important Questions

The result, ACCELERATION $= v^2/R$, brings two questions:

I. How can a moving thing have an acceleration and yet neither go any faster nor get any nearer the center?

---

[3] If you take two lines making an angle X, and turn each line through 90°, you turn the whole pattern through 90° and the two lines in their new position will still make angle X.

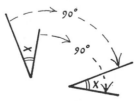

FIG. 21-8.

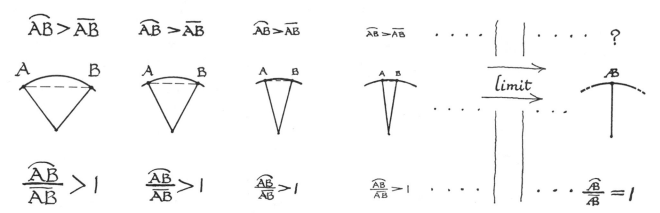

FIG. 21-9. As we proceed from larger arc to smaller and smaller arcs of the same circle, we proceed from large chord to smaller and smaller chords, but the chord grows more and more nearly equal to the arc.

$$IN\ the\ limit$$
$$\frac{arc}{chord} = 1$$

(If you disbelieve this, and claim that the disagreement between arc and chord remains unchanged and is only *disguised* by moving A and B closer, examine the following case (see Fig. 21-10): Choose some size of AB, then change to chord ab half as long, but blow up the new picture to double scale, so that the new *chord* a'b' returns to the full length that you chose originally for AB. Now look at the new chord a'b'. Is it nearer to its arc? Note that the blowing-up does not itself alter the relative proportions of a chord and its arc—it does not change the angles, but merely acts like a magnifying-glass.)

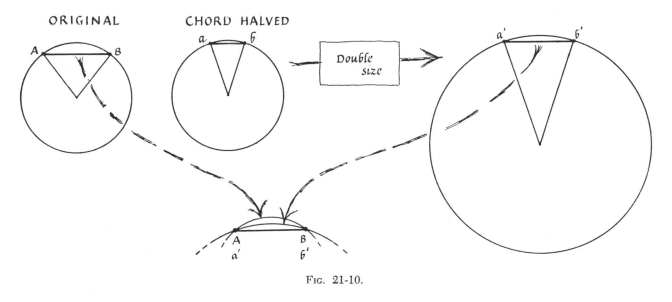

FIG. 21-10.

II. Does this kind of acceleration need a force, according to $F = M \cdot a$, just like a speeding-up acceleration along the path? Does a force $M \cdot v^2/R$ act on a mass M moving around a circle?

Real Forces Needed?

Both questions express real difficulties which kept mankind from jumping to the explanation of planetary orbits at once. Question II is answered by experiment, "Yes every real motion around a circle *does* need a real force, inwards; and Mv^2/R *does* predict the size of that force correctly." To make it move in a circle a body must be pulled or pushed inward by real external agents, such as string or spring or gravity.[4]

[4] People sometimes think that motion in a circle *manufactures* the inward force, provides the inward force needed to maintain itself. A child who *wants* candy does not find his need provides the money to buy it. Some real outside agent such as a rich uncle or an employer must provide the money to buy with—otherwise, no candy. The condition for circular motion is similar. A real outside agent must provide the inward force—otherwise, no curved orbit.

Look for real forces in the following examples:

A. Whirl a stone around with a string. You pull on the string, and the string pulls the stone inward.

FIG. 21-11. WHIRLING STONE ON A STRING
The force on the whirling stone is an *inward* pull, exerted by the string. (The string also pulls the man's hand *outward*—but he is not moving in a circle. He is held in equilibrium by extra forces on his feet.)

The string tugs the stone, gives it some momentum in a new direction, changing the direction of its velocity. Think of the string as giving a series of small tugs: tug to change velocity direction, tug to change it again, tug to change it again, ... all around the circle. If you release the string, the tugging stops, the velocity no longer changes, so the stone *continues steadily along a tangent*. (To say that it "flies off on a tangent," is a misleading description.)

Swing a stone on a string in a horizontal circle, with a spring or a weight providing a measurable inward force. See Fig. 21-12. Any of the experiments sketched can be used as a test of the prediction $F = Mv^2/R$.

B. Watch a "conical pendulum." The bob, which moves in a horizontal circle, is pulled by two real forces, its weight and the string tension.

FIG. 21-13.

(If you measure these forces and add them by vectors you will find they produce a resultant horizontal force inward, towards the center of the bob's orbit. With measurements of dimensions and time of revolution you can test the prediction $a = v^2/R$.)

C. A smooth ball rolls around inside a glass funnel.

D. A steel ball rolls on a horizontal sheet of glass in the field of a magnet pole.

E. The motions of Moon and planets.

FIG. 21-12. TESTS OF $F = Mv^2/R$

FIG. 21-12a. A metal ball, tied to a steel spring by a cord, is whirled steadily in a circle. The spring stretches till it applies a suitable pull, and the length, R, of cord + spring remains constant during whirling. The motion is timed and the predicted value of the inward force, Mv^2/R, is calculated. The force actually exerted by the spring is found by hanging loads on it in a separate experiment. Some indicating device is needed to show how much the spring is stretched during whirling. This arrangement is shown in detail later.

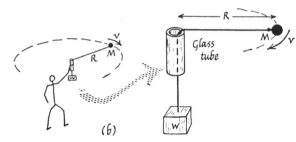

FIG. 21-12b. A metal ball, tied to a cord, is whirled steadily in a circle. The cord runs down a glass tube with smooth open ends and carries a pulling load W at its lower end. By moving the tube around in a tiny circle the experimenter keeps the ball moving around a constant horizontal circle. The motion is timed, and the predicted value of the inward force Mv^2/R is calculated. The force actually exerted by the cord on the ball is its tension, and that is practically equal, but for slight friction, to the pull of the weight of the load W.

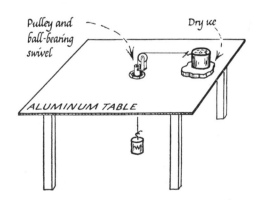

FIG. 21-12c. A practically frictionless version of (b) replaces the ball by a massive block seated on dry ice, coasting on an aluminum table. The cord runs through a hole in the center of the table, and the glass tube is replaced by a small pulley that swivels around the hole on very good bearings.

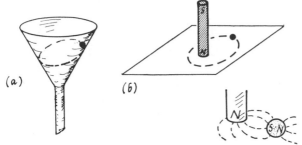

FIG. 21-14.

EXAMPLES OF ORBITAL MOTION WITH CENTRIPETAL FORCE
(a) Steel ball rolling in elliptical orbit in a glass funnel.
(b) Steel ball rolling on a smooth table in the field of
a magnet. (In practice an electromagnet is used in
showing this. It is often placed under the table.) The
magnet's magnetic field magnetizes the ball (? temporarily)
in such a way that it is attracted by the magnet's pole.
Given a suitable push, the ball will then pursue an orbit
around the magnet pole.

PROBLEM 2

Accept for the moment the idea that a body moving in a
circle *must* be pulled inward by a real force (a force supplied
by real agents such as strings, springs, roads). In each of the
following cases, say what agency produces the needed inward
force. (The first answer is given as a specimen.)

(a) Stone whirled on a string, in horizontal circle. (Answer:
"String tension," or "string pulls it.")

(b) (i) Roller-coaster on rails going around sharp corner on
level section of track.

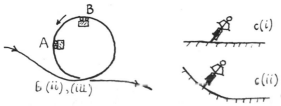

FIG. 21-15 and FIG. 21-16. PROBLEM 2.

(ii) Roller-coaster looping-the-loop at A.
(iii) Roller-coaster looping-the-loop at B.

(c) (i) Bicyclist rounding sharp corner on level road.
(ii) Bicyclist rounding sharp corner on properly banked
track.

(d) Plane flying around a curve.

(e) Negative electron pursuing circular orbit around positive
nucleus of atom.

Acceleration with No Change of Speed

Experiments answer question II above: motion
around a circle *does* need an inward force; and
Mv^2/R *does* predict the size of the needed force.
Now return to question I: how can a thing accelerate towards the center of a circle and yet
neither move faster nor get nearer the center? This
is still puzzling, but it now seems to be more a
matter of wording than physics. The facts are clear;

there are circular motions, and inward forces are
needed to maintain them. These inward forces produce inward momentum changes which swing the
moving object's velocity around, changing its direction without adding to its size. If we like to include
such changes of velocity in our definition of acceleration, $a = \Delta v/\Delta t$, then there *are* accelerations in
circular motion. But *if* we restrict "acceleration" to
its earlier meaning, "going faster and faster," then
there can be *no* "acceleration" in steady motion
around a circle. If we take that restricted view we
must then announce a new set of forces, in addition
to those given by:

"OLD" FORCE = MASS · RATE OF GOING FASTER AND
FASTER

The new forces would be given by:

"NEW" FORCE = MASS · (SPEED)^2/RADIUS

These new forces would have to be real forces, experiment assures us, forces that must be provided
to make a body move in a circle. However, to save
trouble, we avoid the restricted view and use the
name acceleration for *all* kinds of $\Delta v/\Delta t$, because
we find *by experiment*[5] that $F = M \cdot a$ then predicts
the resultant force involved in all cases. On this view
we must undertake two sets of tests and illustrations
of $F = M \cdot a$, one set with little carts pulled along
a track, the other set with objects whirled around
in a circle.

As it goes around a circle, the moving body does
fall *in*, towards the center, in from the tangent it
would otherwise pursue and in again, from the new
tangent line, and in again, and so on, continually
falling in without ever getting nearer the center.
If this seems paradoxical, you may get some comfort
by watching a skater making a small circle on ice—
he leans inward and is falling, yet never falls over.

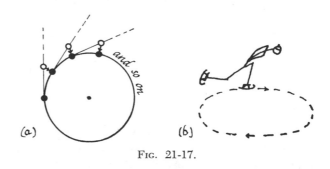

FIG. 21-17.

[5] Galileo really told us this when he said that a projectile's
horizontal and vertical motions are independent of each
other. At the "nose" of its parabola, the vertical acceleration
is perpendicular to the motion and does need a real vertical
force.

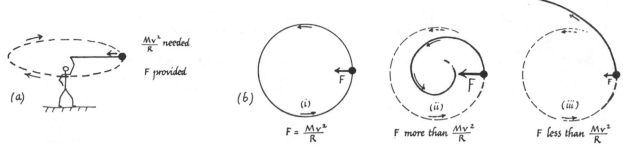

FIG. 21-18. MOTION IN A CIRCLE

(a) Motion in a circle requires changes of *direction of velocity*; and, in that sense, involves a definite, inward $\Delta v/\Delta t$. Experiment shows that this, like "speeding-up acceleration," *needs* a force $M \cdot (\Delta v/\Delta t)$. Geometry shows that the inward $\Delta v/\Delta t$ is given by v^2/R. Then some real agent must provide the needed force $M \cdot v^2/R$.

(b) Of course the pulling agent may fail to provide just the required force Mv^2/R—although strings stretch and rails compress, and friction changes its drag, to adjust the force provided to the need in many cases. The sketches above show what happens when the agent provides (i) the right force, (ii) too big a force, (iii) too small a force. You can arrange these yourself with a stone on a string. In some atomic orbits there may be sudden changes. In planetary motion there are changes which take place smoothly, modifying a circular orbit to an ellipse.

The essential answer to question (I) is this: *the acceleration is perpendicular to the motion,* so it does not increase the speed along the motion. And, added to zero velocity across the motion, the acceleration's contribution just pulls the object in to the standard radius-distance.

Centripetal or Centrifugal Force?

When the force pulls towards the center of a round orbit, producing changes of velocity-*direction* only, we call it a *centripetal* force (from the Latin, meaning a center-seeking force. The opposite of this, a flying-out-from-center force, is called *centrifugal* force).[6] You have often heard that name, but unfortunately it is a misleading term when applied to the moving object. Of course there *is* an outward, centrifugal, pull on the "other fellow" at the center; e.g., on the man who holds the string that whirls a stone. But this gets confused with the force on the moving body, so you will learn good physics most easily if you avoid the phrase "centrifugal force." However, since the idea and phrase are in common use, and are much trusted by many formula-supported engineers, we will discuss it briefly in a later section.

Centripetal Force. Mv²/R

Real agents must provide the needed force which is equal to Mv^2/R, if the mass M is to follow a circular orbit with radius R at fixed speed v. If the

real forces on M produce a greater resultant than this needed force, the body will speed up inward; it will spiral in. If the actual forces are too small it will again fail to follow the circle; it will spiral outward. In many cases the mechanical system adjusts the force to the needed amount, as in the examples below.

EXAMPLES OF CENTRIPETAL FORCE

We shall now discuss in detail some important examples of circular motion ranging from a stone on a string to the modern centrifuge.

Stone whirled on string: the string stretches till its tension is Mv^2/R pulling inward

Roller-coaster looping the loop: the car is pulled by gravity, and it is pushed by the rails. Apart from friction, the rails push perpendicular to their own surface, out from the track. Continue this discussion by answering Problem 3 below.

PROBLEM 3

Suppose a roller-coaster is looping-the-loop as shown in sketch. At A, a force is needed to make the truck move in a circle.

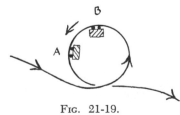

FIG. 21-19.

(a) What direction must that force have?
(b) What provides that force?
(c) What other real force(s) must act on truck at A?

---

[6] These two words have opposite meanings, but their proper pronunciations sound somewhat alike: cen*trip*'t'l and cen*trif*'g'l. To avoid dangerous confusion in discussions, it would be wise to mispronounce them: centri-*pet*'l and centri-*fewg*'l.

(d) What effect must the other force(s) have on the truck's motion?

(*Note:* In answering (d), forget the force discussed in (a) and (b). It must be there, a real force, but it does a special job. Combining by vectors into a resultant force would not be helpful here.)

At B, a certain force is needed to make the truck move in a circle.

(e) What direction must that force have at B?

(f) How is that force provided?

(g) If the truck is moving much slower, a much smaller force is needed. *Why?*

(h) Why may the force *provided* at B be too big? If it is too big, what happens?

Bicycling. When a bicyclist rounds a corner, travelling in a horizontal circle, some real agent must provide an inward, centripetal force. On a rough road, friction does this; it provides a horizontal force, pushing the tires sideways. (On an icy road friction is not available, and the cyclist does not make the corner—he skids straight ahead, thinking, by contrast to his wishes, that he is skidding round a reverse corner.) He leans sideways as he goes round. Leaning does not in itself help friction, but it is necessary, because otherwise the road's push topples him over. If he rides on a tilted cycling track instead of a flat road, he may not need help from sideways friction; the banked track pushes him straight out from its surface with a force which has a vertical component to balance his weight and a horizontal component that provides the needed centripetal force.

Airplanes. A pilot flying round a corner must bank so that air-pressures on his plane push it towards the center of its circular orbit. Instead of just balancing the weight, as they do in steady, straight flight, much greater "lift forces" must be provided through a change of control flaps.[7] The pilot himself must also move in the curve, and thus needs an inward push to make him do it. In the banked plane, his seat pushes him with the required extra force. But what about his blood, circulating to his head (which is pointed towards the center of the banked turn)? The blood must not only make

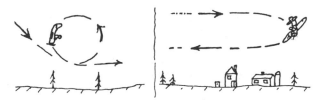

Fig. 21-20a. When a plane loops-the-loop, the pilot's head points towards center. When a plane makes sharp turn, it banks, tilting the pilot over with his head towards center of turning circle.

[7] Gravity, often trivial in a turn, is neglected in this discussion.

Fig. 21-20b. Air exerts forces on wings and body, pushing plane towards center of circle, in loop or turn. (In this sketch, gravity is ignored. In sharp turns it is relatively unimportant.)

the turn, it must move inward ("upward") from his heart to his head. His heart must pump extra hard to feed his brain with an adequate supply. If it cannot pump strongly enough, the blood fails to reach his brain and he faints or blacks out. In ordinary life when you are standing your heart must push a mass m kilograms of blood upward with force $(m \cdot 9.8)$ newtons to keep it moving steadily up against gravity. If the plane flies fast (large v)

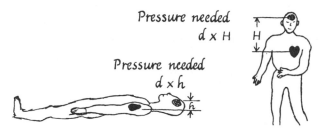

Fig. 21-21. In a man at rest, standing or lying, the heart must provide pressure to pump blood up to brain. In a plane making a turn or loop, the pilot's heart must provide extra pressure for centripetal force, or blood will fail to reach his brain.

around a sharp curve (small R), the needed force from heart to head, $m(v^2/R)$, may be many times bigger than $m \cdot 9.8$. The pilot's heart may be unable to provide the force, mv^2/R, so he may soon black out. If you lie down instead of standing upright, you make much smaller demands on your heart. If, instead of sitting "upright" with his head pointed towards the center of his flying curve the pilot lies down with his body along the curve, he can remain conscious much longer. Fig. 21-22 shows the results of some experiments. The centripetal acceleration of the plane, v^2/R, is expressed as a multiple of "g."

Electrons in atoms. In a later chapter we shall discuss atomic structure and picture electrons whirling around tiny orbits. We assume Newton's Laws apply, and invoke electric attractions to provide mv^2/R.

Cream separators and centrifuges. If we whirl a bottle of liquid around on a string, every chunk of liquid must be given the needed centripetal force mv^2/R, or it cannot make the orbit. This extra inward force must be provided by real pressure-dif-

120

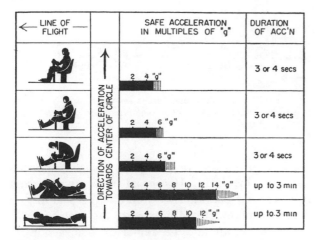

← LINE OF FLIGHT		SAFE ACCELERATION IN MULTIPLES OF "g"	DURATION OF ACC'N
		2 4 "g"	3 or 4 secs
	DIRECTION OF ACCELERATION TOWARDS CENTER OF CIRCLE	2 4 6 "g"	3 or 4 secs
		2 4 6 "g"	3 or 4 secs
		2 4 6 8 10 12 14 "g"	up to 3 min
		2 4 6 8 10 12 "g"	up to 3 min

FIG. 21-22. EFFECT OF POSTURE ON TOLERANCE OF
ACCELERATION
The safe accelerations are those for which the pilot does
not black out. (Data from Ruff)
(From *Nature*, June 10, 1944, vol. 153.)

ferences. The outer end of a chunk must experience bigger pressure than the inner end. Thus there
is a gradation of pressure increasing outwards
much like the vertical gradation of pressure due to
gravity in a liquid at rest, but with fast whirling
these pressures may be far greater. An air bubble,
or anything else less dense than the liquid, would
experience these pressures but would not need so
big an inward force—m being extra small, its needed
mv^2/R would be extra small. It would get more
than it needs and be driven inward. In moving
inward it would accelerate for a short time until it
reached that speed at which fluid friction just balances the available extra force. Then:

INWARD FORCE —	OUTWARD FORCE —	FRICTION DRAG
due to large pressure on outer end	due to smaller pressure on inner end	due to motion inward through liquid

= RESULTANT FORCE
which is the
needed inward
force mv^2/R

That is what happens to cream in a whirling cream
separator. Anything extra dense—a piece of sand
in muddy water, for example—receives the same inward force from surrounding pressure, but needs
more, so fails to make the orbit, spirals out and lands
on the outer edge of the bottle. Take a bottle of
muddy water containing air bubbles. If you stand it
on the table, mud will settle and air bubbles move to
the top with slow steady motions against fluid friction. If you whirl it on a string, mud settles and bubbles rise much faster. Machines called centrifuges

FIG. 21-23. WHIRLING A BOTTLE OF LIQUID ON A CORD

(a)

In each case the actual inward force provided is the
resultant of (END AREA) • (PRESSURE AT OUTER END) *minus*
(END AREA) • (PRESSURE AT INNER END) and these pressures
depend on radii and speeds, but not on contents of sample.
Therefore, the *force provided is the same, for a given
volume, in all cases; but the force needed differs
according to density of sample*—hence the separating action.

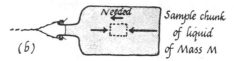

(b) Any specimen chunk of liquid, such as the one
marked in the sketch, is moving in a circle and therefore
requires an inward resultant force to keep it in its
orbit. That force must be provided by difference of
fluid-pressure on its ends. The pressure on the outer
end of the chunk must be greater than the pressure
on its inner end.

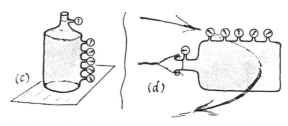

(c) A bottle of liquid *at rest* standing on a table, with
pressure gauges to indicate the gradation of pressure.
(d) A bottle of liquid being whirled in a horizontal circle,
with pressure gauges to indicate the gradation of pressure.
(The vertical effects due to gravity, relatively
unimportant, are ignored here.)

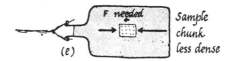

(e) If the immersed chunk is not just a sample of
liquid but has smaller density, it has smaller mass,
therefore needs smaller Mv^2/R. But pressures supply the
same resultant inward push, so sample accelerates IN.

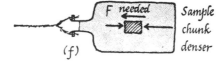

(f) If the sample has greater density than the liquid,
it has greater mass and needs greater Mv^2/R than it gets.
Therefore it accelerates OUT, and settles to what will
be the "bottom" of the bottle when it is placed
upright on the table after whirling.

are used to separate cream, to promote settling of fine sediment, and even to sort out oversized molecules of proteins by their rate of settling, against streamline friction, in water.

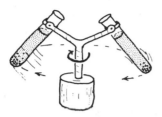

FIG. 21-24. CENTRIFUGE

Square-dancing. When partners "swing" in a square dance, they interlock hands or arms and rotate around a common axis (possibly one or more heels). The partners A and B pull each other inward, the pulls of their arms providing the needed centripetal forces. Even if A has large mass and B small mass, the pulls are equal and opposite (Newton Law III), but the system adjusts its orbit-radii to make these forces just suffice. As a result the dancers rotate around an axis that passes through their common center of gravity.

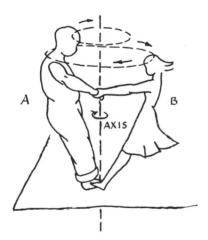

FIG. 21-25. SQUARE-DANCERS

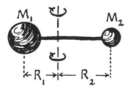

FIG. 21-26.

Here is the algebra of square-dance swinging: Suppose A and B have masses M_1 and M_2 and move around their common axis in circles of radii R_1 and R_2, with speeds v_1 and v_2. One revolution takes time T, the same for A and B. Then $v_1 = 2\pi R_1/T$ and $v_2 = 2\pi R_2/T$ (and $\therefore \dfrac{v_1}{v_2} = \dfrac{R_1}{R_2}$ though this is not needed). The two centripetal forces must be equal and opposite, since they are provided by the action and reaction of the dancers.

$$\therefore \frac{M_1 v_1^2}{R_1} = \frac{M_2 v_2^2}{R_2} \quad \therefore \frac{M_1(2\pi R_1/T)^2}{R_1} = \frac{M_2(2\pi R_2/T)^2}{R_2}$$

$$\therefore M_1 R_1 = M_2 R_2 \text{ (cancelling } (2\pi)^2, \text{ and } T^2, \text{ etc.).}$$

Then MASS · DISTANCE FROM AXIS OF ROTATION is the same for both partners. The massive partner must therefore take a correspondingly smaller distance from the axis. If you consult the rules in physics books for finding the position of the center of gravity of two bodies, you will find that if R_1 and R_2 are measured from an axis through the center of gravity of M_1 and M_2, then $M_1 R_1$ must $= M_2 R_2$, as a property of centers of gravity. And here we find $M_1 R_1 = M_2 R_2$ for the rotating motion. Therefore the rotating pair *must* revolve around their common center of gravity.

This piece of physics is true of square dancers, but it is not important to them. It is also true of the motion of the Moon and Earth, and it is important for an understanding of tides. Astronomers use it for double stars. You will meet other examples of centripetal force in astronomy and in atomic physics.

CENTRIFUGAL FORCE AND THE NOVICE'S HEADACHE-CURE

Motion in a circle needs a real inward force, provided by real external agents. This view of centri*petal* force will help you to deal with all real problems of circular motion. Then what is centri*fugal* force? You often hear of it, may find yourself speaking of it when you whirl something around, and will find books using it to explain things in physics. Here are a variety of opinions on it. You may choose according to your taste.

OPINION I: *"Centrifugal force is a phony force, imagined through a misinterpretation of evidence confusing agent and victim."*

If you whirl a stone on a string, the string-tension pulls your hand outwards (just as it pulls the stone inwards). This is a real centrifugal force on your stationary hand, not on the whirling stone. You feel your hand being pulled outwards, so you say, "I feel the stone and string pulling my hand outwards. That tells me the stone is being pulled outwards, by some centri*fugal* force, and the string is just

FIG. 21-27. ? CENTRIFUGAL FORCE ? OPINIONS I AND II
Some people confuse the inward pull, F_2, on the stone
with the string's outward pull, F_1, on the hand.

transmitting that force." That is where you are mistaken. There is no outward force on the stone. Really the string, in a state of tension, pulls at both its ends. While it pulls your hand outwards it pulls the stone inwards. The only real force *on the stone* is inward, centri*petal*.

Again, suppose you visit one of those amusements at a fair in which people sit on a floor that rotates.

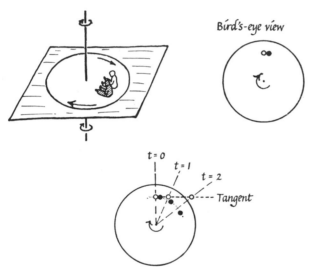

FIG. 21-28. AMUSEMENT AT FAIR OR CIRCUS
The polished floor rotates around a central vertical axis.
If one visitor is allowed to slide, he appears, to the
other visitor to move radially outward, with acceleration.
But an outsider, bird or man, sees him move along tangent.

You and a friend enter the room while the floor is at rest, and sit on the polished floor. Knowing the trick of the performance, you glue yourself to the floor. When the floor begins to spin you note that a mysterious force seems to pull you outward; and, but for the glue, it would make you slide out to the wall. Your friend will slide out to the wall if you do not hold on to him, exerting an inward pull on him. Both of you feel you are struggling against "centrifugal force." But now let a stationary observer take a bird's-eye view from above. Seen from outside the spinning room, you are both moving in a circular orbit, and you both need real *inward* forces to keep you in your orbit. For your friend, the force is the inward

pull you provide; for you it is the pull of the sticky floor on you. Once again, you merely imagined an outward force on your friend because you had to apply a real inward force to him. As the outsider sees, these inward forces are not neutralizing a mysterious outward force, they are *making an inward acceleration*; they are making you move in a curve. The outside observer offers a further comment. When you let go of your friend he then continues along a tangent (if there is no friction). His successive positions along that tangent are farther and farther out from the center of the circle; so, as seen by you spinning with the floor he *seems* to be sliding out along a radius. But really he is just *continuing a straight (tangent) path, a simple example of Newton Law I.*

OPINION II: *"Centrifugal force is a delusion arising from living in the rotating system and trying to forget it."*

The rotating-floor discussion leads straight to this view. To people sitting on the table in a concealing fog—and ignoring its motion—there is an outward field of force, endowing every mass M with an outward force Mv^2/R. Unless some real agent applies an inward force to balance this, any object left alone will seem to slide outward with acceleration v^2/R. Preferring to take a sober view from outside, we say that both the outward field of force and the outward sliding are delusions due to living in a rotating framework and not allowing for its motion.

OPINION III: *The Novice's Headache-Cure*

Here is a good use for centrifugal force. Let us be rude and say, with some truth, that a weak engineering student prefers "Statics," the physics of things at rest (in equilibrium), to the physics of motion. Problems involving acceleration and rotation make his head ache; and he wishes they could be reduced to simple statics problems that he is so good at—forces in bridges and cranes. And they can. Consider, for example, the problem of a pendulum whirling around in a conical motion. The two real forces acting on the bob are its weight and the string tension. These two real forces must add up to a resultant force Mv^2/R inward—otherwise the bob could not continue around the orbit. Here then are two forces W and T which have horizontal resultant Mv^2/R inward. Let us turn this into a statics problem with equilibrium (resultant zero) by adding an extra fictitious force. *What fictitious force must we add to W and T to make zero?* The third force would have to be $-Mv^2/R$, or Mv^2/R *outward*. So

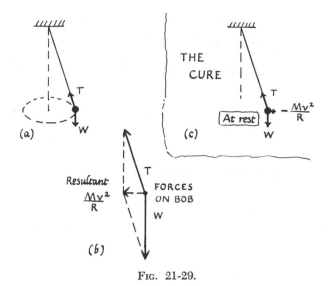

FIG. 21-29.

some teachers say this to the faint-hearted engineer: "Yes, you can turn any problem with circular motion into a statics problem if you *take all the real forces acting on the moving body and ADD a fictitious centrifugal force, Mv²/R outward, and then write an equation stating that these forces (including the fictitious one) have resultant zero.* Solving the equation will give you the same information as the method of making the real forces combine to produce inward acceleration v²/R."

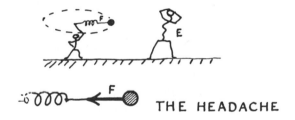

THE HEADACHE

The spring (= agent) provides the real force, F, to make acceleration v²/R

THE CURE

Imaginary force $-\dfrac{M v^2}{R}$ *+ real force F make equilibrium*

FIG. 21-30.

On this view, centrifugal force is a fictitious force, but a useful one, to cure the engineer's headache. It is also used thus in advanced physics, to save trouble—but then it is a sophisticated trick in the hands of skilled craftsmen. As used by most students, it gives the right answer but makes some of the theory harder to understand—how can it help that when it reduces obvious motion to fictitious rest? The trustful user, with his right answer, is confused about the forces: he is not sure which are real or which way they pull. *If you value your understanding of physics, avoid this headache-cure at all costs.* Of course, a *mixture* of this centrifugal headache-cure with centripetal forces will produce utter confusion!

OPINION IV: *Relativity.*

(This opinion sketches some comments from very sophisticated relativity theory. Read it for amusement or for a good moral warning, but do not let it convert you to the engineer's headache-cure for problems. This relativity-view is true, but only within the framework of definitions constructed for it.) Can nothing better be said of centrifugal force? Returning to Opinion II, some scientists ask, "Why is it so wicked to view things from a rotating framework? After all, we live on a spinning Earth. Are the 'centrifugal forces' that arise from our rotating-framework viewpoint really different from other forces, and less real? Who are we to say which is really rotating, ourselves or everything else?" (We are back to Copernicus vs. Ptolemy.) This last question is like the problem of testing Newton's laws in an accelerating railroad coach. By building a tilted room in the coach we could still find the same laws, though we should find "gravity" changed in size and direction. We suspect that we *cannot* distinguish between the effect of acceleration and a change of gravity—Einstein built General Relativity theory on an elaboration of that "cannot."

Relativity theory starts with an axiomatic statement, that we cannot tell which is moving, ourselves or "the other fellow," that there is no such thing as *absolute* motion. If that is so, "absolute space" is meaningless; it should not be used, and cannot be needed, in science. In that case, the working geometry of "space" must be such that we discover the same physics whether we think we are moving or "the other fellow" is. And that makes us modify the simple geometry of space and motion that Euclid assumed and Galileo and Newton used. For constant velocities, we have many experimental failures to distinguish absolute motion even with the help of light-signals, so we feel justified in accepting the Relativity principle and its modified geometry. In practical life, the modifications are not noticeable, and they only affect experiments noticeably when very high speeds are involved, as they are in astronomy and in atomic physics. Extending the Relativity attitude to accelerated motion we assume that a local observer will find the effects of acceleration indistinguishable from a local change of gravity; and thus we decide that gravitational fields can be treated as local changes of geometry in space-&-time. This is Einstein's Principle of Equivalence. Though the viewpoint is entirely new, its practical form shows only small deviations from Newton's law of gravitation.

Extending this idea to rotation, we suggest that a local observer cannot distinguish between the effects of a rotating framework and a local change of gravity, if he is moving with that frame. In that case centrifugal force tugging outward would be just as real to him on his spinning floor as an extra, horizontal pull of gravity. Then, to a bug in a centrifuge, centrifugal force-fields should appear just like real gravitational fields, only some thousands of times as strong as ordinary gravity. To the bug, gravity would take on a new direction—he would quite forget about its old direction—and it would be enormously stronger. This General Relativity view has proved useful in coordinating thinking and successful in making predictions; and so far we have not observed anything inconsistent with it. In this way, centrifugal force has grown to be respectable. When we want to test the effects of large gravitational fields, unattainable on Earth, we think we may use a centrifuge instead.

The general principle of equivalence forbids us to call the motions of the Earth absolute. It therefore leads to a new mechanics and geometry that will predict the same effects whether the Earth spins and moves around the Sun, or the stars and Sun move around us. On General Relativity theory, a rotating universe would produce "centrifugal forces" at a stationary Earth; so tests of a spinning Earth, with a Foucault pendulum[8] or equatorial changes of "g," could not distinguish between the two causes: Earth spinning or everything-else-spinning. Faced with the old question, "Is Copernicus right and Ptolemy wrong?" we must demur at Galileo's cocksure insistence and say, "Both views *may* well be equally true, though one is a simpler description for practical thinking and working." Here is Hegel's development: thesis . . . antithesis . . . synthesis.

[8] See p. 42.

OPINION ON THE FOUR OPINIONS?

Make your own choice. However, for problems and experiments in this course, you are advised to use only centri*petal* force.

PROBLEMS FOR CHAPTER 11

1. In the following problems assume: (i) that the centripetal acceleration is v^2/R, and (ii) that F = Ma applies to this motion. (*Remember that whenever F = Ma is involved the force must be in newtons or poundals, if the mass is in kilograms or pounds.*)

(a) A 2.00-kilogram stone is whirled in a horizontal circle on a level frictionless table by means of a string. The string is 4.0 meters long, so the circle has radius 4.0 meters. The stone moves with speed 7.0 meters/sec around its orbit. Calculate, with a word or two of explanation:
  (i) the stone's acceleration. (Leave it in factors.)
  (ii) the tension in the string (state the units of your answer).

(b) Suppose the string just breaks under the tension calculated in (a). What is its breaking force in *kilograms-weight?*

(c) As in (a), a 2.00-kilogram stone is whirled in a circle by a rope 4.0 meters long but with such a speed that it makes 5 revolutions in 2 seconds.
  (i) Calculate the orbital speed. (Leave answer in factors, keeping π as π.)
  (ii) Calculate the tension in the rope. State the units of the answer. (A rough answer will suffice. You may take $\pi^2 \approx 10$.)

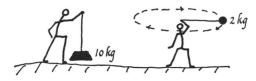

FIG. 21-42. PROBLEM 6

2. A certain kind of string can just carry 10.0 kilograms hung on it vertically, but breaks with the smallest increase of load.

(a) What is its breaking force in kilograms-weight?
(b) What is its breaking force in newtons?
  A piece of this string 1.00 meter long is used to whirl a 2.0-kilogram stone in a horizontal circle faster and faster till the string breaks.
(c) Calculate the stone's maximum *speed* around its orbit, giving a short explanation of your calculation.

3. A lacrosse player running with the ball weaves his stick to and fro in front of him as he runs. He does not move it in a straight line from side to side, but swings it in a curve that is concave towards him. Explain how this motion prevents the ball from falling out of the net of his stick.

4. A plane flying 600 ft/sec (410 miles/hr) is pursuing a small slow plane flying 300 ft/sec. The slow plane turns around and runs away by flying in a horizontal semicircle; and the fast plane tries to follow. The pilot in each plane can just stand an acceleration of 5g.

(a) Calculate the radius of the smallest semicircle the pilot of the slow plane can safely make at 300 ft/sec.

(b) How long (roughly) will the slow plane take to turn around its *semicircle*?

(c) Calculate the radius of the smallest circle the pursuer can safely make.

(d) Where will the pursuer be when the slow plane has finished its semicircle? (Mark its path on a sketch.)

5. ALTERNATIVE DERIVATION OF $a = v^2/R$ (Newton's method).

As the body moves around the circle from A to B, *treat it as a falling body with constant acceleration downward.* Then in time t it falls distance h with acceleration a, from rest.

(a) Write an equation for a in terms of h and t assuming a is constant.

(b) Using a geometrical property of chords of a circle, write an equation expressing h in terms of other measurements in the diagram.

(c) Substitute the expression for h in the equation of (a).

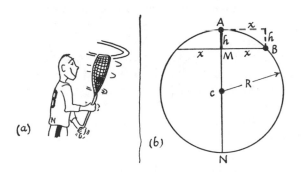

FIG. 21-43.
(a) PROBLEM 7   (b) PROBLEM 9

(d) Now imagine B is moved closer and closer to A. As $B \to A$, the horizontal distance $x \to$ arc $\overset{\frown}{AB}$. And, as $B \to A$, the chord MN $\to$ full diameter, $2R$. Make these changes in your expression for acceleration.

6. CENTRIFUGING (Estimate roughly throughout this problem).

(a) A centrifuge whirls its test-tube in a circle of average radius 1 ft, at 5000 revolutions/*minute*, thus placing the contents in a force-field of strength many times "g." How many times "g"?

(b) A sample of muddy water contains particles about the size of blood-corpuscles (diameter 10^{-5} meter). If it is standing in a vertical test-tube, the particles fall at constant speed about $\frac{1}{4}$ inch/minute. So a sample about 4 inches high clears in about $\frac{1}{4}$ hour. (The particles do not settle completely to the bottom. Brownian-motion diffusion keeps some in suspension.) How long would the same sample take to clear in the centrifuge of (a)?

(c) Protein molecules (several hundred times smaller in diameter than the mud of (b), but large compared with the simpler molecules of, say, salt or air) fall about 300,000 times slower in water than the mud particles of (b). How long would a 4-inch tube of a suspension of such protein molecules in water take to clear in the centrifuge of (a)?

(d) Without the use of the centrifuge, the suspension of protein would never clear. Why?

(e) From the rates of clearing, the diameters of the particles involved can be compared, if their densities are known. (Friction drag on a small sphere $\propto$ radius and $\propto$ velocity.) The mud specks of (b) can be measured with a microscope. The protein molecules are invisible. "Chemical" measurements (e.g., osmotic pressure) show the protein molecules are about 10^6 times as massive as a hydrogen atom. What important *atomic* information could the centrifuging yield?

CHAPTER 12 · ISAAC NEWTON (1642-1727)

~~~~~~

"If I have seen farther than others, it is by standing on the shoulders of giants."
—An old saying, quoted by Newton.

~~~~~~

PRELIMINARY PROBLEM FOR CHAPTER 22

★ PROBLEM 1. NEWTON'S FIRST TEST OF
GRAVITATION.

Newton did not explain why apples fall. (Calling the cause of weight "gravitation," from the Latin and French for "heavy," does not explain. Saying "the Earth attracts the apple" does put the blame on the Earth and not the sky, but beyond that gives no further information about the attraction.) However, faced with the problem "what keeps the Moon and planets moving in their orbits?" Newton was able to suggest an "explanation" in the sense that the same property of Nature is involved in these motions as in the already-well-known apple fall. "Explaining" therefore just meant linking together under a common cause—but this linking is very useful for further predicting and for simplifying our understanding of Nature.

Operating first on the Moon, Newton calculated its actual acceleration, v^2/R. This is much smaller than the ordinary value of "g," 32.2 ft/sec². Therefore, the Moon cannot merely be "falling under gravity" unless Earth's gravity out there is much smaller, diluted by distance. Newton tried using a simple form of distance-dilution, an inverse square law: he assumed that at twice the distance "g" would be ¼ as big, at ten times the distance 1/100 as big, &c.

Using the data below, repeat Newton's test, calculating (accurately—see note below*):

(a) The Moon's *actual* acceleration, in ft/sec², assuming $a = v^2/R$.

(b) The *expected* value of "g" at the Moon, in ft/sec², assuming the ordinary value of "g" is diluted according to an inverse square law. (Assume that the Earth pulls an apple as if all the material of the Earth is concentrated at the center, one Earth's radius from the apple.)

Since the answers are asked for in FEET/sec², it is important to express *distances* in *feet* and *times* in *seconds* before using them in any calculation. However, you may write in the conversion factors and postpone the multiplying-out if you like, till you find you have to do it. A mixture of miles, hours, feet, seconds, and hope will lead to frustration.

Data: Radius of Earth = 3957 miles.
           Radius of Moon's orbit is 60.3 times radius of
           Earth
           1 month is 27.3 days. (This is the Moon's absolute
            period, relative to the stars.)
           1 mile is 5280 feet.
           "g" for the apple is 32.2 feet/sec².

---

* This was a crucial calculation, a great test of a theoretical idea. Therefore, in trying it for yourself you need to do the arithmetic correctly, or it is a complete waste of time. The two answers to (a) and (b) which are to be compared should each be calculated to three significant figures (which will require five decimal places in ft/sec²). First show your answers to (a) and (b) in factors without any cancelling, and then reduce to final decimals.

Newton's Life and Work

Newton was born in the year Galileo died. Even as a boy he enjoyed experimenting—like Galileo and Tycho he made ingenious toys such as water mills,[1] and he even measured the "force" of the wind in a great storm by noting how far he could jump *with* the wind and *against* it. He went to grammar school, doing poorly at first in its principal subject, Latin, then showing unusual promise in mathematics. His uncle, who acted as informal guardian, sent him to the university when he was 19. There, at Cambridge, he devoured a book on logic and Kepler's treatise on optics (so fast that he found no use in attending the lectures on that subject). He read Euclid's geometry, finding it childishly easy; then started on Descartes' geometry. He had to work hard to master this, but he threw himself heart and soul into the study of mathematics. Soon he was doing original work. While still a student he discovered the binomial theorem,[2] and by the time he was 21 he had started to develop his work on infinite series and "fluxions"—the beginning of calculus. He was too absorbed in his work, or too shy, to publish his discoveries—this curious absent-mindedness or dislike of public argument lasted

---

[1] Early play with bricks, toys, stoves, and bathtubs provides a store of experimental knowledge which we call common sense. When we say "common sense tells us that" we are often appealing to such knowledge—though sometimes to prejudice or tradition instead.

[2] Binomial theorem:

$$(1 + x)^n = 1 + \frac{n}{1}x + \frac{n(n-1)}{1 \cdot 2}x^2 + \frac{n(n-1)(n-2)}{1 \cdot 2 \cdot 3}x^3$$

+ . . . etc. If n is a positive whole number, the series ends after $(n + 1)$ terms. If not, the series is infinite and x must not exceed 1 for the statement to be true. When x is much smaller than 1, we can say $(1 + x)^n \approx 1 + nx$, because the later terms are so much smaller. This provides some useful approximations such as $(1 + x)^3 \approx 1 + 3x$, if x is small

$$\sqrt{1 + x} \approx 1 + \tfrac{1}{2}x, \text{ if } x \text{ is small.}$$

Note how this shows that an error of $y\%$ in a factor Y makes an error of $3y\%$ in Y^3, or of $\tfrac{1}{2}y\%$ in $\sqrt{Y}$.

Examples: When a solid is heated and its linear dimensions expand by 0.02%, its volume expands by 0.06%. When a clock's pendulum expands in length from winter to summer by 0.02%, its time of swing increases only 0.01%.

through his life. He was also interested in astronomy, observing Moon halos and a comet. Later he was to design and make his own telescopes. He took his bachelor's degree, continuing to work on mathematics and optics, helping the professor of mathematics with original suggestions. In the next two years he turned his attention to the solar system. He began to think of common gravity extending to the Moon and holding the Moon in its orbit, like a string holding a whirling stone. He arrived at the formula for centripetal acceleration, $a = v^2/R$, which he needed to test his idea for the Moon— he found this before Huygens published his version of it. Then he extended the idea to the planets and imagined them held in their orbits by gravitational pulls from the Sun. Thus he guessed at universal gravitation: attractions between Earth and apple, Earth and Moon, Sun and Mars, Sun and Earth, . . . Kepler's Law III told him that these attractive forces must decrease at greater distances, and that the attraction must vary as $1/\text{DISTANCE}^2$. He was already making his great discoveries. When asked how he made his discoveries, Newton replied that he did it by thinking about them.[3] This seems to have been his way: quiet, steady thinking, uninterrupted brooding. This is probably the way in which much of the world's greatest thinking is done. Genius is not solely patience or "an infinite capacity for taking trouble"; yet patience and perseverance must go with great ability and insight for the latter to bear their fullest fruit.

In his use of the Moon's motion to test his new theory Newton met a serious difficulty, so he put that work on one side, shelving it for some years, and threw himself heart and soul into optics, buying prisms, grinding lenses, delighting in his experiments on the spectrum.

By the time he was 24 he had laid the foundations of his greatest discoveries: differential and integral calculus, gravitation, theory of light and color; but he had revealed little of his results. Then his mathematics teacher consulted him about a new discovery in mathematics that was being discussed. Newton made the surprising reply that he had worked this out himself, among some other things, some years earlier. The papers he fetched showed he had gone further and solved a more general form of the problem. This made such an impression that, when the professor retired shortly after, Newton was elected,

at 26, to one of the most distinguished mathematical professorships in Europe. In his new post, he lectured on optics but he still did not publish his work on calculus. He was invited to give a discourse to the newly-formed Royal Society of London on his reflecting telescope. The members were delighted with his talk and proceeded to elect him a Fellow. In later lectures he expounded to them his discoveries concerning color.

It was then, after six years, that he returned to his work on astronomy. He could now carry through his test on the Moon's motion with delightful success. Yet for a dozen years more he worked on in silence. Meanwhile Kepler's Laws were begging for explanation. The idea of gravitation was in the air. The members of the new and flourishing Royal Society were arguing about it. They could prove that some inverse-square-law force would account for circular orbits with Kepler's Law III, but they found elliptical orbits too difficult. One of them appealed to Newton for help, and he calmly explained that he had already solved the problem, that he knew, and could prove, that *an inverse-square-law-force would require planets to obey all three of Kepler's Laws!*

Then came a time of writing and publishing (not always willingly) and extending his work on mechanics and astronomy and the mathematics that went with it. His friends in the Royal Society persuaded him to publish his theory of the solar system. The book he then produced was far more than that; it was the world's greatest treatise on mechanics: definitions, laws, theorems, all beautifully set forth and then applied to a general gravitational theory of the solar system, with explanations, examples, and far reaching predictions—a magnificent structure of knowledge. This was the *Principia*, "The Mathematical Principles of Natural Philosophy."

An appreciative people made him a Member of Parliament and later Master of the Mint. This was a way of rewarding him financially as well as honoring him. Newton took the work seriously and did some work on metallurgy, extending his earlier interest in chemistry.[4] Through most of his life he seems to have wanted to be a man of importance and property. This appointment, as well as his election to Parliament, fulfilled the wish in some measure. When he was 61 he was made President of the Royal Society and held that very distinguished office for twenty-four years, the rest of his life. When

[3] His remark was "by always thinking unto them. I keep the subject constantly before me and wait till the first dawnings open slowly by little and little into a full and clear light."

[4] At Cambridge he had carried out long chemical investigations, recording a wealth of detail, but even he was hampered by the chaotic state of chemical knowledge and thinking at that time.

he was 65, he was knighted, becoming Sir Isaac Newton. The people of his country, and their neighbors far and wide, realized that they had more than a great man, a very great man, and they did their best to honor him and provide for him in a time when scientists were only just coming to their own.

When he died at 85, Newton left books on laws of motion, gravitation, astronomy and mathematics, and optics, in addition to many writings on religion. (He was a religious man with devout, though rather unorthodox views—perhaps something like those of Unitarians today. Having unravelled astronomy with magnificent success, he hoped to do the same for religion.) He had raised astronomy to an entirely new plane in science, bringing it to order by a general explanation in terms of laws which he laid down and tested. Since we are concerned here with the growth of astronomy, we now return to an account of Newton's work in that field.

Laws of Motion

To bring all the heavens into one explanatory scheme, Newton needed rules for motion. He found in Galileo's writing clear statements about force and motion, with a less clear understanding of the nature of mass. He re-stated Galileo's findings in clear usable form, carefully defining his terms. In his *Principia* he published his statements in the form of two laws, after he had been using them in his work for some time, and he added a third law supported by his own experiments.[5]

In this he was a great law-maker, a codifier, the Moses of Physics. Of course, Moses stated laws with an entirely different attitude, codifying heavenly commands to be obeyed by man, while Newton was codifying Nature's ways and man's interpretation of them. Yet both were extracting, codifying, and teaching. Moses did not invent all the laws and and rules he set forth; he gathered them together,

---

[5] He probably formulated his Law III, when he found the need for it in developing mechanics systematically—others had already put forward the general idea. He tested it by experiment thus: he allowed moving pendulum bobs to collide, and measured their velocities before and after collision. He calculated the *changes* of momentum and found these were equal and opposite. Arguing from his own Law II, he concluded the forces involved must be equal and opposite. That was an honest test of Law III, but he had to make a *housekeeping-assumption* about masses: that when several are put together the total mass is got by adding—one mass does not shield another. (Attempts to "prove" Law III by pulling two spring balances against each other, or by allowing ingeniously for the effects of air-friction on his pendulums. He gives a detailed account in his *Principia*, available in good translations, and quoted in Magie's *Source Book in Physics*.

edited them so to speak, and put them clearly so that the people could understand them. As a great lawmaker he was a great teacher. Newton, like Moses, was a great teacher, though shy and modest, almost teaching himself rather than others. Much of his writing was done originally to make things clear for himself, but it served when published to clarify, for the scientists of his day and of ages to come, what had been difficult and unclear.

Nature and Nature's Laws lay hid in Night.
God said, "Let Newton be"; and all was Light.
—Alexander Pope.

Newton wrote his laws to be clear, not pompous; though, since he disliked argumentative criticism by amateurs, he kept his mathematics tough and elegant. He used Latin for his *Principia* because it was the universal language among scholars. When he wrote in English (e.g., his book on Optics) he used the English of his day, and where that now seems pompous or obscure it is because vocabulary and usage have changed with time. If he were writing his laws now he would want them well worded and would write them in English, avoiding the involved phrases that lawyers love. Here are the three laws in their original Latin from the *Principia*, followed by versions in ordinary English.

LEX I

Corpus omne perseverare in statu suo quiescendi vel movendi uniformiter in directum, nisi quatenus a viribus impressis cogitur statum illum mutare.

LEX II

Mutationem motus proportionalem esse vi motrici impressae et fieri secundum lineam rectam qua vis illa imprimitur.

LEX III

Actioni contrariam semper et aequalem esse reactionem: sive corporum duorum actiones in se mutuo semper esse aequales et in partes contrarias dirigi.

Newton added comments and explanations, in Latin, after each law. Using modern terminology we translate the laws thus:

LAW I

Every body remains at rest or moves with constant velocity (in a straight line) unless compelled to change its velocity by a resultant force acting on it.

LAW II

When a force acts on a body, RATE OF CHANGE OF MOMENTUM, (Mv), varies directly as the RESULTANT FORCE; *and the change takes place in the direction of that force.*

OR

The product MASS · ACCELERATION varies directly as the RESULTANT FORCE, and *the acceleration is in the direction of that force.*

LAW III

To every action there is an equal and opposite reaction.

Or, when any two bodies interact, the force exerted *by the first body on the second* is equal and opposite to the force exerted *by the second body on the first.*

Note that in the English version of Law II, Newton's form which uses momentum is given first. This is still the best, most general, form today; but the second form, using acceleration, is often used in elementary teaching because it seems simpler. Here is the reverse change from the momentum version to $F \propto Ma$:

RATE OF CHANGE OF MOMENTUM varies as FORCE

$$\therefore \frac{\Delta(Mv)}{\Delta t} \propto F \quad \therefore \frac{(Mv_2 - Mv_1)}{\Delta t} \propto F$$

$$\therefore \frac{M(v_2 - v_1)}{\Delta t} \propto F \text{ if } M \text{ remains constant.}$$

$$\therefore M(\Delta v/\Delta t) \propto F \quad \therefore Ma \propto F \quad \text{or, } F \propto Ma.$$

Or, in one stride with Newton's own invention, calculus:

$$\frac{d(Mv)}{dt} = M\frac{dv}{dt} = Ma \text{ if } M \text{ is constant.}$$

We assumed above that the moving mass remains constant. When the mass is changing the first version, $\Delta(Mv)/\Delta t \propto F$, gives correct predictions, and this is the version that Newton chose. He must have seen that it would apply to a moving object gaining mass (e.g., a truck with rain falling into it). He could not have foreseen its extension to modern Relativity where we still use it to define force in terms of mass which increases with velocity—an increase that is undetectable *except at very high speeds.*

Earlier views

Motion had been worrying scientists for some time.

Leonardo da Vinci (150 years before Newton), had stated, probably just as guesses copied from still earlier writers:

(1) If a force moves a body in a given time over a certain distance, the same force will move half the body in the same time through twice the distance.
(2) Or: the same force will move half the body through the same distance in half the time.
(3) Or: half the force will move half the body through the same distance in the same time.

Descartes (some 40 years before Newton), stated:

(1) All bodies strive with all their might to stay as they are.
(2) A moving body tends to keep the same speed and direction. (He gave a theological reason.)
(3) The measure of a body's force is its mass (not clearly defined) and its speed.

Query: How many of these early views on motion seem true to Nature? (At least one of them seems quite wrong.)

Out of such earlier statements, together with Galileo's books and his own thinking, Newton produced his three Laws of Motion. Today we trust them to predict many kinds of motion: a ball rolling downhill, a rocket starting, planets in their orbits, and even streams of electrons deflected by fields. Einstein has added a modification, but the original summaries are very close to the real behavior of Nature.

Newton's Laws: Natural Truths or Definitions?

Like any modern scientist, Newton tried to give clear definitions of velocity, momentum, and force. In science, a definition is not an experimental fact, or a risky assumption or a speculative idea. It is a piece of dictionary-work explaining as precisely as possible how we are going to use a word or phrase or even an idea. For example, we define "acceleration" as "$\Delta v/\Delta t$" and thereafter whenever we say "acceleration" we mean, definitely, GAIN OF VELOCITY/TIME TAKEN, or RATE-OF-GAIN-OF-VELOCITY, and we do not mean something else such as $\Delta v/\Delta s$, or something vague such as "going faster." We define "gravitational field-strength at a point" as "FORCE, due to gravity, on UNIT MASS, placed at that point"; and that is both a description of what we mean by field-strength and a definite statement showing how it is measured.

Newton's Laws were clear, powerful rules, based on observation of mechanical behavior, meant to be used to predict other cases of mechanical behavior. However, they were not merely statements extracted from experiment. They incorporated definitions and descriptions of words and ideas such as mass and momentum; and they provided a consistent scheme of prediction *in terms of those definitions*. Definitions often take part in theory like that. For example, two centuries after Newton the science of thermodynamics was developed to produce amazing predictions of heat-properties, in terms of a temperature scale. But this scale had to be a particular scale *defined in the scheme of thermodynamics itself*. We find disagreement when we compare this scale (the Kelvin scale) with other scales such as that given by the mercury thermometer, or that given by the gas thermometer. Yet we cannot say one scale is "right" and another "wrong." Each scale is defined clearly and unambiguously, and all three are equally right as ways of giving a precise measure to man's vague sense of hotness. But when we have a particular use in mind one scale may be best, and when we have a particular theory we may be restricted to the scale whose definition is incorporated in the theory. In using thermodynamic theory and its predictions we *must* use the Kelvin scale of temperature. Fortunately the Kelvin scale is almost the same as the common mercury thermometer's scale, so we can put the predictions to practical uses.

This close interweaving of experiment and definition to make the structure of a theory is characteristic of modern science. So if you examine Newton's Laws critically you may come to the conclusion that Law I merely explains what is meant by force—in fact, defines the nature of force—and that Law II defines the measurement of either force or mass. Then perhaps the laws are just our own inventions: a put-up job? That is going too far. Both laws refer to real nature as revealed by experiment and there is a solid core of fact in them—though that may be hard to disentangle logically from the definitions involved.

Two centuries after Newton stated his laws, further doubts and difficulties began to appear. Newton had adopted "Galilean relativity." In his mechanics it did not matter whether the observer was at rest or moving with constant velocity. Yet Newton thought absolute space must be identifiable by rotational effects. (If the Earth were at rest with all the heavens spinning around it once a day, should we observe the Earth's bulge, differences of gravity, Foucault's pendulum slewing around?) So Newton wrote of absolute motion, forces producing absolute accelerations—not just accelerations relative to some

moving frame. Yet where is the fixed, unmoving frame in space? Earth, Sun, stars, may all be moving. *Is* there a real fixed frame? If we cannot find one, we may be foolish to talk in terms of it. Out of these doubts grew Relativity theory.[6] In your present studies of mechanics you would be wise to forget these doubts and take Newton's Laws as good simple working rules. In using them for solving problems, remember that they are at best carefully worded summaries of agreed definitions and experimental knowledge. They are not sacred laws to be cited to make things true! All they do is remind us how to interpret past experiments, and how to predict what will happen in some future ones. At the same time they teach us useful ideas or concepts, such as mass and momentum.

Newton and Planetary Motion

Newton formulated his laws so that he could use them. Turning then to astronomical problems, he at once had the answer to the problem which had misled the Greeks and had puzzled Kepler and even Galileo: "What keeps the Moon and the planets moving along their orbits?" Crystal spheres, natural circular motions, rotating spokes of magnetic influence, vortices—had all been suggested. Newton saw that these explanations provided an agency where none is needed. To keep a planet moving, no force is needed (Law I). Left alone it would just go straight on forever. But a force *is* needed to pull its path into a curved orbit, away from the no-force straight line. How big an inward force is needed? And what agency provides *that* force? These were Newton's new questions. If Law II holds for this motion, the required force F must be MASS · ACCELERATION. But what *is* the acceleration for this motion in an orbit? Newton attacked uniform motion in circular orbits first. (The orbits of the Moon and most planets are nearly circular.) He arrived at the same result as others who were working on the same problem around that time: ACCELERATION, directed inward along the radius, $= v^2/R$, where v is the orbital speed and R the orbit radius. (See Ch. 11 for the derivation of this from the definition of acceleration. Geometry is involved, but no knowledge of force or mass. Newton's proof was an unusual one, treating the moving body as a projectile and each short piece of the circle's circumference as a "nose" of the projectile's parabola.) Then the force must be

---

[6] Simple forms of Einstein's Relativity deny any fixed framework of "space" but more thoroughgoing discussions of General Relativity still refer their predictions (such as the slow slewing of Mercury's orbit) to a framework made by the most distant stars.

Mv^2/R, inward along the radius. So the Moon, moving steadily around a circular orbit, is always accelerating inward toward the Earth, yet never getting any nearer. You may think of it as falling in to the circle from a straight tangent path, reaching the orbit just in time to start falling in enough to reach the next part of the orbit, still without getting any nearer. If this seems paradoxical, remember that any projectile thrown in a parabola is *accelerating* down ("g") just as much as ever at the nose or vertex of its parabola—yet at that point it is not *moving* upward or downward, not getting any nearer the ground. Thus, it is possible to have acceleration without any velocity at the moment in its direction. The Moon's orbit is a series of "noses," so to speak.

Now came Newton's guess to explain what provides this force. He suggested that the agency which makes bodies near the Earth fall may also pull the Moon in to follow its orbit. There is a story, probably true, that he was thinking about this problem while sitting in an orchard, and an apple falling on his head suggested the answer. We call such pulls "gravity," a word which just means heaviness or implies some connection with weight, and this name in itself is no explanation. The commoner

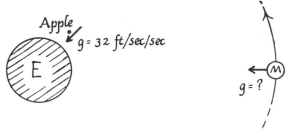

FIG. 22-1. EARTH'S GRAVITY

word, weight, is better for most purposes. Newton suggested that it is the Moon's *weight* that keeps it in its orbit. If the Moon were very near the Earth's surface its weight would give it an acceleration, "g," about 32 ft/sec²—the same for the Moon as for an apple, except for the bulkiness of the Moon spoiling such an experiment. Does the Moon out in its real orbit have this acceleration? Is v^2/R for the Moon about 32 ft/sec²? The Moon takes a month of 27.3 days to travel once around its orbit relative to the fixed stars. Newton knew the Moon's orbit-radius, R, was 60 Earth's radii, 60r. He had a rough value for the Earth's radius, r. So he could calculate the Moon's speed, v, in the form (circumference $2\pi R$)/(time T, one month) and thence the Moon's actual acceleration v^2/R. The answer is *far* smaller than 32 ft/sec². If "gravity" is responsible, it must

be much weaker out at the Moon's orbit. Newton guessed at a simple dilution-rule, the inverse-square law. This is the rule governing the thinning out of light (and radio) and sound, and the force due to a magnetic pole or electric charge—the rule for anything that spreads out in straight lines from some "source" without getting absorbed.[7] He had got a hint by working backwards from Kepler's third law! He tried this inverse-square-law rule. For a Moon at 60 Earth's radii, instead of an apple at only one Earth's radius from the center of the Earth, the pull should be reduced in the proportion of $1:\frac{1}{60^2}$ or $1:\frac{1}{3600}$. Then the Moon's acceleration should be not 32 ft/sec² but $\frac{32}{3600}$ ft/sec². You can easily compute v^2/R for the Moon and will find that it comes out very close to that "predicted value." Think what a joy it would have been if you had discovered this agreement for the first time. It is a successful test of a combination of $F = Ma$ and $a = v^2/R$ and the inverse-square law of gravitation. You would have made a first check of a tremendous theory; a great discovery would have been yours.

Yet Newton himself, eager but far-seeing, was not entirely happy about this test. He mysteriously put the whole calculation aside for some years. He was probably worried about calculating the attraction of a bulky sphere like the Earth. He diluted "g" by a factor 60², but this dilution, 1 to $\frac{1}{60^2}$, assumes that the body near the Earth, with "g" = 32 ft/sec², is *one Earth's radius* from the attracting Earth. Does the great, round Earth attract an apple as if the whole attracting mass were 4000 miles below the surface, at its center? Some of the Earth's mass is very near the apple and must pull very hard (according to an inverse-square law, which for the moment is being assumed). Some of the Earth is 8000 miles away and must pull very little.

[7] Suppose a small sprayer acts as a butter-gun and projects a fine spray of specks of butter from its muzzle, in straight lines in a wide cone. If the cone just covers one slice of bread

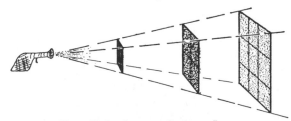

FIG. 22-2. INVERSE-SQUARE LAW

held one foot from the muzzle, it will cover four slices at 2 feet or nine slices at 3 feet, &c.; therefore, for specimen slices of bread placed 1 foot, 2 feet, 3 feet, . . . from the muzzle, the *thickness of buttering* will be in the proportion $1:\frac{1}{4}:\frac{1}{9}$. . . . This is the "inverse-square law of buttering."

132

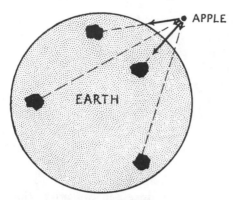

FIG. 22-3. NEWTON'S PROBLEM: ATTRACTION OF SPHERE
Apple being attracted by various parts of the Earth.
(Four specimen blocks of Earth are shown.) What
will be the resultant, for whole Earth?

Other parts of it pull in slanting directions. What
is the resultant of all these pulls? Here was a
very difficult mathematical problem; the adding up
of an infinite number of small, different pulls. It is
easily done by integral calculus, but that fine mathematical tool was just being constructed. Newton
himself invented it, for this and other purposes in
his work, simultaneously with the German mathematician Leibniz. He shelved his Moon calculation till he knew from calculus that *a solid sphere
attracting with inverse-square-law forces attracts as
if its mass were all at its center.* "No sooner had
Newton proved this superb theorem—and we know
from his own words that he had no expectation of
so beautiful a result till it emerged from his mathematical investigation—than all the mechanism of
the universe at once lay spread before him."[8] Then
he could attack the Moon's motion again and test,
in one single calculation, his laws of motion, his
formula, v^2/R, and his great guess of inverse-square-
law gravity as the cause of the Moon's round orbit.
This time he was overjoyed with the calculation.
The agreement was excellent; the needed force *is*
provided by diluted gravity. He had explained the
mystery of the Moon's motion.

Newton's Explanation

In one sense Newton had *explained* the mystery,
by saying that gravity holds the Moon in its orbit.
In another sense, he had not explained anything.
He had produced no explanation whatever of gravity, no "reason why" to explain the mystery of
gravity itself. He had only shown that the same
agency causes, or is concerned in, both an apple's
fall and the Moon's motion. This finding of general

[8] J. W. L. Glaisher, on the bicentenary of the *Principia*,
1887, quoted in Dampier, *A History of Science* (Cambridge
University Press, 1952), p. 153.

causes common to several things is called "explaining" in science. If you are disappointed, reflect that
this process does simplify our picture of Nature and
note that in common speech "to explain" means to
make things clearer. It also means to give reasons
for, and in Newton's work, as in most science, the
basic reasons or *first causes* do not emerge; but
things that seemed to have different causes are
shown to be related. Thus, while we learn more
about Nature by finding these common explanations, the basic questions of how the Universe
started, and why things in it behave just so, remain
unanswered.

Universal Gravitation

So, plain gravity—or rather diluted gravity—is
the tether that holds the Moon in its orbit. How
about the planets? Does a similar force hold them
in their orbits? Since they move round the Sun
rather than round the Earth, the force must be a
Sun-pull not an Earth-pull. To deal with this, Newton guessed at universal gravitation: a universal
set of mutual attractions, with an inverse-square
law. He said he reasoned it must be that, by working backwards from Kepler's Law III.

Every piece of matter in the Universe, he guessed,
attracts every other piece of matter. He knew,
from the Myth-and-Symbol experiment, that the
weights of bodies (Earth-pulls on them) are proportional to their masses. Thus the Earth's attraction varies as the mass of the victim. So the attraction exerted by Earth of mass M_1 on mass M_2
varies as M_2. If the attraction is mutual (Law III),
symmetry vouches for a factor M_1 to match M_2. For
changes with distance, the Moon's motion had given
a single successful test of inverse-square attraction.

UNIVERSAL GRAVITATION

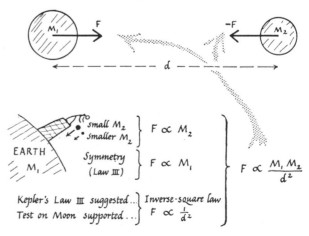

FIG. 22-4.

So Newton tried a factor $1/d^2$ in his general law. Thus he formulated his Law of Gravitation:

$$F \propto \frac{M_1 M_2}{d^2} \qquad \text{or } F = (\text{constant}) \cdot \frac{M_1 M_2}{d^2}$$

$$\text{or } F = G \frac{M_1 M_2}{d^2}$$

where G is a *universal* constant, M_1 and M_2 are the two masses and d is the distance between them. F is the force with which each pulls the other. Note that G, the universal constant, is quite a different physical quantity from g, the local value of "gravity."[9]

Would this general law account for planetary motion? Newton showed it would. He proved that such an attraction would make a planet move in an ellipse with the Sun in one focus. It was easy (for him) to show that Kepler's other two laws also followed from his guess of universal gravitation. (These laws hold for a sun attracting alone. Then we must add the effects of other planets attracting the moving planet. In the solar system those are trivial compared with the pull of the massive Sun— trivial but far from negligible in the precise accounting of modern astronomy.)

Thus Newton carried his simple idea for the Moon far out to the whole planetary system. He assumed that every piece of matter attracts every other piece, with a force that varies directly as each of the two masses and inversely as the square of the distance between them; and from that he deduced the whole detailed motion of the solar system, already codified in laws that had been tested by precise measurements over two centuries. Satellites of planets formed a similar scheme. Even comets followed the same discipline. All these motions were linked with the gravity that was well known on Earth. Newton explained the heavenly system in a single rational scheme.

This is so great an achievement that you should see for yourself how Newton derived Kepler's three laws—and then put them to further use. The first proof, the ellipse, requires either calculus, which Newton devised, or cumbersome geometry (which Newton had to provide as well, to convince those who distrusted his new calculus). So, with great

---

[9] For the case of the Earth and the apple, M_1 is the Earth's mass and M_2 the apple's; and their distance apart is the Earth's radius, r. So the apple's *weight*, $M_2 g$, must be $G M_1 M_2/r^2$. This shows the relation between g and G:

$$g = G M_1/r^2$$

ACCELERATION (OR FIELD-STRENGTH), g,

$$= \frac{\text{GRAVITATION CONSTANT} \cdot \text{MASS OF EARTH}}{(\text{RADIUS OF EARTH})^2}$$

regret, we shall not give it here. We shall now derive Kepler's Law III, and then Law II—equal areas in equal times. Law II follows from *any* attractive force-law, provided the force always pulls *directly from planet to Sun*; but Kepler's first and third laws fit only with inverse-square attraction.

Kepler's Law III

To deduce Kepler's third law, Newton had merely to combine his laws of motion with his law of universal gravitation. For elliptical orbits, calculus is needed to average the radius and to deal with the planet's varying speed, but the same law then follows.

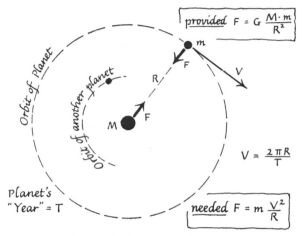

FIG. 22-5. PLANETARY MOTION

For circular orbits, suppose a planet of mass m moves with speed v in a circle of radius R around a Sun of mass M. This motion requires an inward resultant force on the planet, mv^2/R to produce its centripetal acceleration v^2/R (see Ch. 11). Assume that gravitational attraction between sun and planet just provides this needed force. Then

$$G \frac{Mm}{d^2} \text{ must} = \frac{mv^2}{R}$$

and distance d between m and M = orbit-radius, R.

$$\text{But } v = \frac{\text{circumference}}{\text{time of revolution}} = \frac{2\pi R}{T}$$

where T is the time of one revolution

$$\therefore G \frac{Mm}{R^2} = m \frac{(2\pi R/T)^2}{R} \qquad \therefore G \frac{Mm}{R^2} = \frac{4\pi^2 m R^2}{T^2 R}$$

To look for Kepler's Law III, collect all R's and T's on one side; move everything else to the other.

$$\therefore \frac{R^3}{T^2} = \frac{GM}{4\pi^2}$$

Now change to another planet, with different orbit radius R' and time of revolution T', then the new value of $(R')^3/(T')^2$ will again be $GM/4\pi^2$; and *this has the same value* for all such planets, since G is a universal constant and M is the mass of the sun, which is the same whatever the planet. Thus R^3/T^2 should be the *same for all planets* owned by the sun, in agreement with Kepler's third law. For another system, such as Jupiter's moons, M will be different (this time the mass of Jupiter) and R^3/T^2 will have a different value, the same for all the moons.

The planet's mass, m, cancels out. Several planets of different masses could all pursue the same orbit with the same motion. You might have foreseen that—it is the Myth and Symbol story on a celestial scale.

With any other law of force than the inverse-square law, R^3/T^2 would *not* be the same for all planets. An inverse-cube law, for example, would make R^4/T^2 the same for all; then values of R^3/T^2 would be proportional to $(1/R)$, and *not* the same for all planets. In fact, as Kepler found, they are all the same. The inverse-square law is the right one.

Calculus predicts Law III for elliptical orbits too, where R is now the average of the planet's greatest and least distances from the Sun.

Kepler's Law II

Here is a crude derivation, due to Newton. We use Newton's Law II, CHANGE OF MOMENTUM $= F \cdot \Delta t$. Then changes of mv are vectors, along the direction of F, proportional to F.

First suppose we have a planet moving under *zero* force. We can still draw a radius to it from a

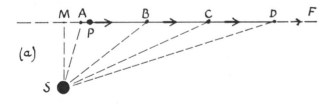

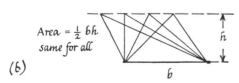

FIG. 22-6.

(a) MOTION OF A PLANET WITH NO ATTRACTION. Planet P moves in a straight line with constant speed. SP sweeps out equal areas in equal times.

(b) THE PROPERTY OF TRIANGLES USED HERE. All triangles on the same base and with the same height have the same area. Another version: If triangles have the same base and their vertices lie on a line parallel to the base, their areas are equal.

non-attracting sun, S ! The planet, P, moves with fixed speed in a straight line, AF (Newton Law I). Mark the distances travelled by the planet in equal intervals of time: AB, BC, CD, . . . etc. Since the speed is constant, AB = BC = . . . , etc. Consider the areas swept out by the radius SP. How do the following triangles compare, SAB, SBC, SCD? All these triangles have the same height, SM, and equal bases, AB, BC, CD. Therefore all their areas are equal: the spoke from S sweeps out equal areas in equal times. This simple motion obeys Kepler's second law.

Now suppose the planet P moves in an orbit because the Sun pulls it inward along the radius PS. But, to simplify the geometry, suppose the attraction only acts in sudden big tugs, for very short times, leaving the planet free to travel in a straight line betweenwhiles. Then it will follow a path such as that shown in Fig. 22-7. Suppose it travels AB, BC,

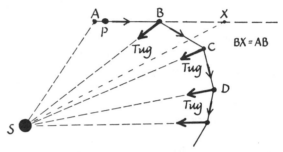

FIG. 22-7. MOTION OF A PLANET WITH TUGS OF ATTRACTION Without tug at B, P would move on to X.

CD, etc., *in equal times*, the inward tugs occurring abruptly at B, at C, at D, etc. The planet moves steadily along AB; then, acted on by a brief tug at B, along BS, it changes its velocity abruptly and moves (with new speed) along BC. Except for the tug at B the planet would have continued straight on, as in the simple case discussed above. On this continuation, mark the point X an equal distance ahead, making AB and BX equal. But for the tug at B, the planet would have travelled AB and BX in equal times, and the radius from S would have swept out equal triangles, SAB and SBX. But in fact the planet reaches C instead of X. Does this change spoil the equality of areas? If the planet travels to C, the two areas are SAB and SBC. Are these equal? To change the motion from along AB to along BC, the tug at B pulls straight towards the Sun, along BS. This tug gives the planet some inward momentum along BS, which must combine with the planet's previous momentum to make the planet move along BC. The planet's previous momentum was along AB.

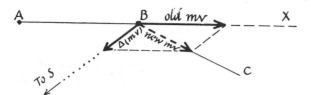

FIG. 22-8. ENLARGED SKETCH OF MOMENTUM-CHANGE AT B

Therefore,

| original momentum along AB | + | gain of momentum inward along BS | must = | new momentum along BC. |

Newton's Law II reminds us that momentum is a vector. So the adding must be done *by vector addition* (see Fig. 22-8). As the planet's mass is constant, we may cancel it all through and use velocities thus:

| velocity along AB | + | gain of velocity along BS | must = | velocity along BC |

Let us use the *actual distance* AB to represent the planet's velocity along AB. Then, BX must also represent this velocity and BC must represent the planet's new velocity along BC (since all these are distances travelled in equal times). Using this scale, we make a vector diagram (see Fig. 22-9) express-

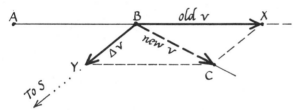

FIG. 22-9. COPY OF FIG. 22-8, WITH VECTORS TO SHOW VELOCITIES AT B
Scale has been chosen so that AB or BX represents original velocity along AB, before tug acts at B.

ing the equation above. Use BX (= AB) for the original velocity before the tug. Use BC for the velocity after. The change of velocity must be shown by some vector BY along BS straight towards S. Complete the parallelogram, with BC the diagonal giving the resultant. Because this is a parallelogram *the side XC is parallel to BY*, so C lies on a line parallel to BS.

Now look at the triangles SBC and SBX, in Fig. 22-10. They have the same base, BS, and lie between the same parallels, BS and XC, so they have equal areas. Therefore, area of SBC = area SBX, which = area SBA. Therefore, the triangles SBA and SBC have equal areas. By a similar argument,

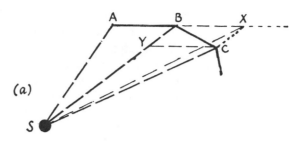

(a)

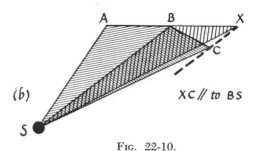

(b)   XC // to BS

FIG. 22-10.
(a) FIG. 22-7, redrawn with C in its proper place on XC parallel to BY or BS.
(b) FIG. 22-10a, redrawn with equal-area triangles shaded.

the triangles SBC and SCD have equal areas, so all the triangle areas are equal, and Kepler's second law does hold for *this* motion. This argument only holds if all the tugs come from the same point S. If we now make the tugs more frequent (but correspondingly smaller) we have an orbit like Fig. 22-11, nearer to a smooth curve, and Kepler's law still holds, *provided the tugs are directed straight from planet to Sun.* If we make the tugs still more frequent, we approach the limit of a continual force, with an orbit that is a smooth curve. The argument extends to this limit, so Kepler's Law II holds for a smooth curved orbit.

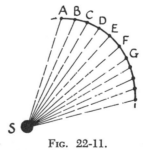

FIG. 22-11.
The equal time-intervals from A to B to C . . . are much shorter. Orbit is nearer to a smooth curve. With a smooth-curve orbit, the segments swept out in equal times may each be regarded as a bunch of small triangles like those here. So the segments must have equal areas.

Kepler's Law II and Rotational-momentum.

Newton deduced Kepler's second law from his mechanical assumptions. The inverse-square law is not necessary; any *central* attraction, directed straight to Sun as center, will require this law of equal areas.

In advanced mechanics, this is treated as a case of *Conservation of Rotational-momentum (or Angular Momentum)*. What is rotational-momentum,[10] and why are we sure it is conserved? Here is a short account, too rough to be convincing but intended to give you an idea of this fundamental conservation law. For motion along a straight track, we have quantities such as: DISTANCE (s), VELOCITY (v), ACCELERATING FORCE (F), . . . , and laws or relations such as $F \cdot \Delta t = \Delta(Mv)$, . . . , and Principles such as *Conservation of Momentum*. When we have a body that is only spinning without moving along, we can apply Newton's Laws to each moving part of it and produce an equivalent scheme. Instead of DISTANCE MOVED we have ANGLE TURNED (in revolutions or in radians). Instead of VELOCITY we have RATE-OF-ROTATION (in r.p.m. or in radians/sec). Instead of FORCE we have the MOMENT OF FORCE or TORQUE, which is FORCE · ARM—the common-sense agent to make a thing spin faster and faster. Then instead of

FORCE · TIME = GAIN OF MOMENTUM,

Newton's Law II gives

TORQUE · TIME = GAIN OF ROTATIONAL-MOMENTUM.

Guess at rotational-momentum and you will probably guess right: just as TORQUE is FORCE · ARM ($F \cdot r$), so ROTATIONAL-MOMENTUM is

---

[10] For a spinning body we might call this "spin-momentum" or simply "spin"—which we do for an electron—but, for a planet swinging around a remote Sun, we use the general name "rotational-momentum." This name applies to spinning objects and objects revolving in an orbit.

MOMENTUM · ARM, ($Mv \cdot r$). Multiply F and Mv each by the arm from some chosen axis, and the new rotational version of Newton's Law II is just as true. In all this, the arm is the *perpendicular arm* from the axis to the line of the vector force or momentum.

Now suppose two bodies that are not spinning collide and exert forces on each other so that one body is left rotating. The FORCES are equal and opposite (Newton Law III); and the ARM from any chosen axis is the same arm for both the forces. Therefore, the TORQUES around that axis on the two bodies are equal and opposite during the collision. Therefore, the ROTATIONAL-MOMENTUM gained by one body must be just equal and opposite to the ROTATIONAL-MOMENTUM gained by the other. Therefore, the total ROTATIONAL-MOMENTUM generated must be zero. If one body develops rotation, the other body must also develop a counter rotation around the same axis. In *any* collision or other interaction: *rotational-momentum is conserved*—it can only be exchanged without loss, or created in equal and opposite amounts.

Therefore, an isolated spinning body (e.g., a skater whirling on one toe) cannot change its rotational-momentum. It cannot change the total of all the $Mv \cdot r$ products of its parts. Suppose it shrinks (skater draws in his outstretched arms). The "r" decreases for the parts drawn in, and if the total rotational-momentum stays constant, Mv must increase: *the body must spin faster*. Watch a skater. Whether he wishes to or not, he spins faster when he draws in his arms or curls up an extended leg.

(a) STRAIGHT FORWARD MOTION | IN ANY COLLISION:

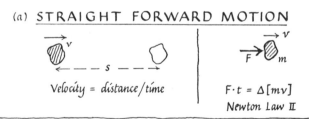

Velocity = distance/time | $F \cdot t = \Delta[mv]$ Newton Law II

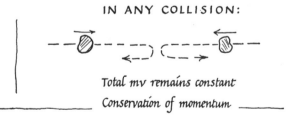

Total mv remains constant Conservation of momentum

(b) ROTATIONAL MOTION

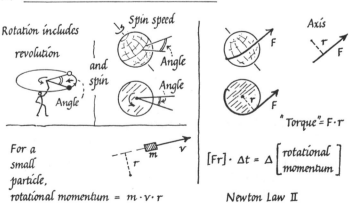

Rotation includes revolution and spin

For a small particle, rotational momentum = $m \cdot v \cdot r$ | "Torque" = $F \cdot r$ $[Fr] \cdot \Delta t = \Delta \begin{bmatrix} \text{rotational} \\ \text{momentum} \end{bmatrix}$ Newton Law II | IN ANY COLLISION:

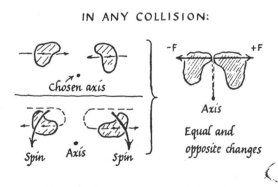

Total rotational momentum remains constant

FIG. 22-12. ROTATION

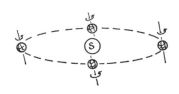

FIG. 22-13. A SPINNING BALL, with no torque acting on it from outside, keeps its rotational-momentum constant in size *and* direction. (e.g. spinning Earth).

"An *isolated* spinning body cannot change its rotational-momentum." Apply that to the spinning Earth. Apply that to a man spinning on a *frictionless* piano stool. Turn yourself into an "isolated spinning body" as follows: pivot yourself by standing on one heel so that you can spin around several times before friction stops you. (Better still, stand or sit on a

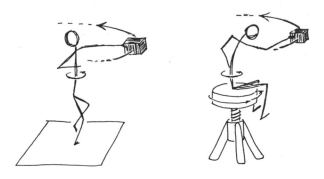

FIG. 22-14. MAN ON "FRICTIONLESS" PIVOT changes *speed* of spin when he pulls load in nearer to axis.

turntable that rotates freely.) Hold a massive book at arm's length. Now, as you spin, bend your arm and pull the book suddenly toward you. Notice what happens to your speed. Here is rotational-momentum being conserved. But here is Kepler's Law II: the book is a "planet" drawn nearer by you as "Sun" as it revolves in its orbit. (Your own great mass is hopelessly involved in this rough test, so you will not observe Kepler's Law accurately.)

For a real planet, if the Sun pulls along the direct spoke, *that pull has no torque around an axis through the Sun; so that pull cannot change the planet's rotational-momentum around the Sun.* A real planet has rotational-momentum $Mv \cdot r$ around the Sun, where r is *not* Kepler's spoke but the *perpendicular* arm from the Sun to the line of velocity v (the orbit tangent). As the planet moves nearer the Sun, r decreases and, to keep Mvr constant, v must increase in the same proportion. Sup-

pose that in a very short time t the planet moves along a short bit of arc s with velocity $v = s/t$. Then the planet's rotational-momentum around the Sun is $M(s/t) \cdot r$, or Msr/t. But $s \cdot r$ is BASE · PERPENDICULAR HEIGHT for the thin triangle swept out by the

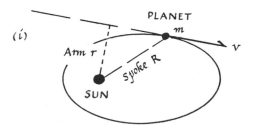

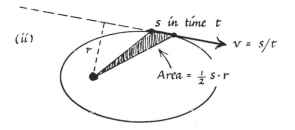

FIG. 22-15. ROTATIONAL-MOMENTUM OF PLANET Equals $mvr = m \cdot (s/t) \cdot r = m$ (TWICE AREA SWEPT OUT)/TIME.

spoke in this time. That is twice the triangle's area. Therefore,

planet's ROTATIONAL-MOMENTUM
$$= \frac{(\text{MASS},M)(\text{twice the AREA SWEPT OUT by spoke})}{\text{TIME}, t}$$

For direct pull by Sun, $\frac{\text{AREA SWEPT OUT}}{\text{TIME}}$ cannot change: the RATE-OF-SWEEPING-OUT-AREA is constant; Kepler's Law II must hold. Conversely, when Kepler discovered his equal-areas-in-equal-times law, he was showing that the only force on the planet is directed straight towards the Sun, and that there is no other kind of force such as a friction-drag due to viscous æther.

Of all the great conservation-rules of mechanics—holding the total constant for mass, momentum, etc.—the Conservation of Rotational-momentum is as universal as any. In atomic physics we shorten it misleadingly to Conservation of "Spin" and expect it to hold through thick and thin, even in violent interchanges between atomic particles and radiation.

Fruitful Theory

Newton formed his theory: he framed his laws as starting points by intelligent guessing helped by hints from experimental knowledge; then he deduced consequences, such as Kepler's laws, then tested those deductions against experiment. In the case of Kepler's laws, the experiments were already done. Tycho's observations had provided rigorous tests; so when Newton deduced them his experimental tests of theory were ready for him. With these successes, there seemed little doubt that the theory was "right." By this stage it seemed worth more than the separate facts that went into it. It gave a simple general meaning to planetary behavior by linking it with the familiar facts of falling bodies, and it offered hopes of many further predictions. Newton, armed with powerful mathematical methods and guided by an uncanny insight, applied his theory to a variety of problems in his book the *Principia*. Some of these are described below.

1. *Masses of Sun and Earth*. Newton calculated the mass of the Sun in terms of the Earth's mass. (The Earth's mass itself was not known and could not be estimated without some terrestrial measurements like Cavendish's. See Ch. 13.) His calculation can be carried out as follows. Subscripts $_s$ and $_e$ and $_m$ refer to Sun and Earth and Moon.

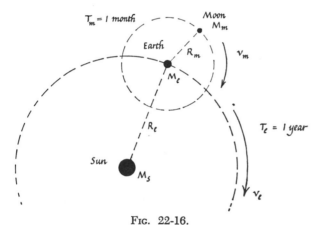

Fig. 22-16.
CALCULATING THE RATIO OF SUN'S MASS TO EARTH'S M_s/M_e, by using the motion of the Moon. The Moon's mass M_m cancels out. For each motion, Earth-around-Sun and Moon-around-Earth, just write an equation stating that the needed force Mv^2/R is provided by gravitation.

For the Earth's motion around the Sun in its yearly orbit,

$$G\frac{M_sM_e}{R_e^2} = M_e\frac{V_e^2}{R_e} = M_e\frac{4\pi^2R_e^2}{R_eT_e^2}$$

$$\therefore M_s = \frac{4\pi^2}{G}\left[\frac{R_e^3}{T_e^2}\right] \text{ Note that the Earth's mass, } M_e, \text{ cancels}$$

For the Moon's motion around the Earth in its monthly orbit,

$$G\frac{M_eM_m}{R_m^2} = M_m\frac{v_m^2}{R_m} = M_m\frac{4\pi^2R_m^2}{R_mT_m^2}$$

$$\therefore M_e = \frac{4\pi^2}{G}\left[\frac{R_m^3}{T_m^2}\right] \text{ Again, the Moon's mass, } M_m, \text{ cancels}$$

Therefore, dividing one equation by the other

$$\frac{M_s}{M_e} = \left[\frac{R_e^3/T_e^2}{R_m^3/T_m^2}\right] = \frac{R_e^3}{R_m^3}\frac{T_m^2}{T_e^2}$$

$$= \left[\frac{\text{DISTANCE OF SUN}}{\text{DISTANCE OF MOON}}\right]^3\left[\frac{1\text{ month}}{1\text{ year}}\right]^2$$

With the known values of these times and orbit radii, the ratio of the Sun's mass M_s to the Earth's mass M_e can be calculated.

2. *Masses of planets*. Newton could make similar estimates for the mass of Jupiter or any other planet with a satellite, in terms of the Earth's or Sun's mass. (Our own Moon has no obvious satellite yet; so its mass, which cancels out in the first equation applied to it, seems difficult to find.)

3. *"g" at equator*. Since the Earth spins, an object should seem to weigh less at the equator than at the pole, because some of its weight must be used to provide the needed centripetal force to keep it moving in a circle with the Earth's surface. An object at rest on a weighing scale must be *pushed up* by the scale less than it is pulled down by gravity (its weight). Therefore, the object's *push down* on the scale (which is what the scale indicates), must be less than its weight by the small centripetal force mv^2/R. The Earth's gravitational field-strength must *seem* less. Newton calculated this small modification of "g," which is now observed, together with the effects of the Earth's spheroidal shape.

4. *The Earth's bulge*. Newton calculated the bulging shape of the Earth, arguing as follows. Suppose the Earth was spinning with its present motion when it was a pasty half-liquid mass. What shape would it take? To answer this, consider a pipe of water running through a *spherical* Earth from the North Pole to the center and out to the Equator. If this were filled with water, just to the Earth's surface at the North Pole, where would the water-level be in the equatorial branch of the pipe? The water pressure at the bottom of the polar pipe is due to the weight of the water in that pipe; and this pressure pushes around the elbow at the bottom and out along the other branch, trying to push the column of water up that branch. The weight of water in that branch

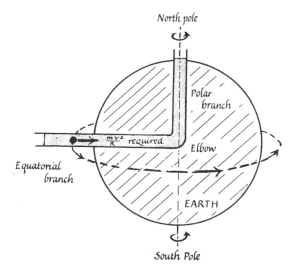

FIG. 22-17. To ESTIMATE THE BULGE OF A SPINNING EARTH, imagine a pipe of water running from North Pole to center and out to equator. Calculate the extra height of water in equatorial branch needed to provide mv^2/R forces for spin. This gives extra radius of bulge for a pasty Earth congealing while spinning.

pulls it down. But these two forces on water in the *equatorial branch* must be unequal. They must differ by enough to provide an inward centripetal force to act on the water in that pipe, which is being carried around with the spinning Earth. The weight of the water in that branch must *exceed* the upward push from the water at the elbow by the amount needed for mv^2/R forces. Therefore, the water column in this pipe must be taller than that in the polar pipe. The equatorial pipe must extend out beyond the Earth's surface to carry the extra height of water. Newton calculated the extra height and

found that 14 miles would be required. He argued that the Earth at an early pasty stage would bulge out about this distance. This bulge had not yet been observed. A short time later, measurements of the Earth confirmed the prediction. Jupiter shows a more marked elliptical shape.

5. *Precession.* Newton explained the precession of the equinoxes thus: the axis of spinning Earth is made to slew around in a cone by the pulls of Sun and Moon on its equatorial bulge. The Earth's axis is tilted and not perpendicular to the ecliptic plane of the Earth's orbit; so the equatorial bulge is subject to unsymmetrical gravitational pulls by Sun and Moon. Here we shall describe the effect of the Sun's pulls. The Sun would pull a *spherical* Earth evenly, as if all the Earth's mass were at its center. The resultant pull would run along the line joining centers of Earth and Sun, whether the Earth spins or not (Fig. 22-19a). A spheroid with an equatorial bulge is subject to small extra pulls on the bulge (Fig. 22-19b). These pulls are uneven, the largest pull being on the portion of bulge nearest the Sun (Fig. 22-19c). These small extra pulls are equivalent to an average pull on the whole bulge, along the line of centers, *plus* a small residual force, *f*, which tries to rock the spin-axis (Fig. 22-19c). Since the Earth's axis is tilted, this force *f* is a slanting one, off center, with greatest slants at midsummer and midwinter. When such a slanting pull acts on any *spinning* body it does *not* succeed in rocking the body over in the expected way. Instead, it produces a very curious motion, called precession, which you have seen when a spinning top leans over while spinning fast. The pull of gravity on the leaning top does not

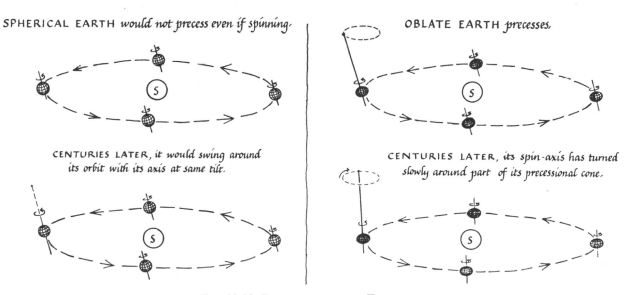

FIG. 22-18. PRECESSION OF THE EQUINOXES

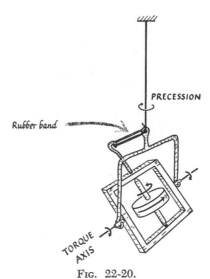

FIG. 22-19. PRECESSION
(a) The Sun would pull a spherical Earth with a
central pull along line joining centers,
whether the Earth spins or not.
(b) The Sun exerts extra unequal
pulls on the bulge of oblate Earth.

(c)

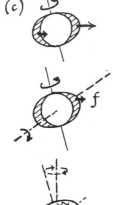

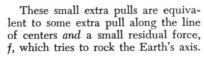

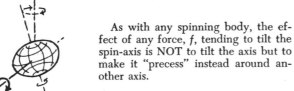

(c) The Sun pulls the nearer part of the bulge harder than it pulls the remoter part.

These small extra pulls are equivalent to some extra pull along the line of centers *and* a small residual force, *f*, which tries to rock the Earth's axis.

As with any spinning body, the effect of any force, *f*, tending to tilt the spin-axis is NOT to tilt the axis but to make it "precess" instead around another axis.

make it fall over, but makes the spin-axis slew slowly round a cone. Newton showed that the attractions of the Sun, and even more the Moon, would make the Earth's axis precess round a cone of angle 23½° taking some 26,000 years to complete a cycle (Fig. 22-19c). Here was a deduction of the precession that was observed by the Greeks and expressed more simply by Copernicus, but entirely unexplained until Newton's day. It had seemed such a strange motion that men must have had little hope of finding a simple explanation. Yet Newton showed that it is just one more result of universal gravitation: the spinning Earth is made to precess like an unbalanced top.

Demonstration Experiment

Fig. 22-20 shows an experiment to illustrate precession of the Earth. A frame carrying a rapidly spinning flywheel is hung on a long thread. The thread and frame enable the wheel to twist freely about a vertical axis or about a horizontal axis (in the frame, perpendicular to the vertical axis). Left

FIG. 22-20.
EXPERIMENT TO ILLUSTRATE THE PRECESSION OF THE EARTH

alone, the tilted spinning wheel continues to spin without other motion. Then a rubber band is attached so that it pulls the frame with a force that tries to rock the spinning wheel about the horizontal axis; the wheel does not obey the rocking force in the obvious way, but instead precesses around the vertical thread as axis.

Explaining precession of a gyroscope

The Earth, and a spinning top, and a "mysterious gyroscope" all precess in the same way, for the same reason (Fig. 22-21). Precession looks mysterious, but it is just a complicated example of Newton's Laws applied to the parts of a spinning body. With no off-center force, the body retains its spin, unchanging in size or direction. With an off-center force making a rocking torque, we can compound rotational-momentum changes as vectors to show that the axis will precess. The geometry of precession is given in standard texts.

Here is a simpler explanation that shows precession as a straight case of Newton's Law II. Fig. 22-22 shows a large wheel with massive rim hung by a cord, PQ, and precessing. Consider the momentum of a small chunk of rim at the top, A. It is moving forward; but the wheel's weight, rocking the wheel over a little, moves A to the right, giving A a little momentum to the right. This momentum is added to A's main momentum forward: so after a short time A has momentum in a skew direction, *forward and to the right*. Similarly B, at the bottom, develops skew momentum *backwards and to the left*. Then, for A and B to have these momenta, *the wheel must have twisted round the vertical axis* (i.e., precessed). Here is a hint of the mechanism of preces-

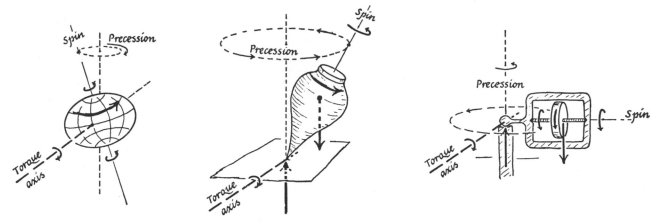

FIG. 22-21. PRECESSION
The Earth, a spinning top, and a "mysterious gyroscope" all precess in the same way, for the same reason. In the sketches above "torque axis" means the axis around which the tilting force tries to rock the spinning object.

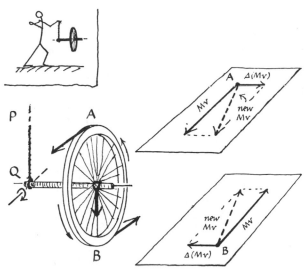

FIG. 22-22. PRECESSION,
treated as a case of Newton's Law II. The sketch shows a large bicycle wheel with massive rim, precessing.

sion but it is hard to visualize the extension of this discussion to the rest of the moving rim.

6. *Moon's motion.* The Moon suffers many disturbances from a uniform circular motion. For one thing, it moves in a Kepler ellipse, as any satellite may, with the Earth in one focus. But that orbit is upset by small variations[11] in the *Sun's* attraction. The Moon is nearer the Sun at new moon than two weeks later at full Moon, and that makes changes of attraction that hurry and slow it in the course of a month. This effect is exaggerated when the Sun is nearer in our winter, so there is also a yearly varia-

[11] The Sun exerts a strong, almost constant, attraction on the Moon, and the average of this is just sufficient to make the Moon accompany the Earth in its yearly orbit: a solar Myth-and-Symbol.

tion of speed. In addition, the changes in Sun-pull make the Moon's orbit change its ellipticity; they tilt the Moon's orbit up and down a little; they slew the orbit-plane slowly around; and they make the oval orbit revolve in its own plane. Newton predicted these effects on the Moon's motion, making estimates of their size where he could. Some effects had long ago been observed; some were actually being sorted out—and Newton begged the Astronomer Royal for the measurements; some were not observed till afterwards. For the revolution of the elliptical orbit in its own plane, measured as 3° a month, Newton's first calculations predicted only 1½°. For years after Newton, mathematicians wrestled with the problem, trying to explain the disagreement—they even tried modifying the inverse-square law with an inverse 3rd power term. Then one of them found that some terms of Newton's algebra had been neglected unjustifiably and that with them theory agreed with experiment. Still later, it was found from Newton's papers that Newton himself discovered the mistake and obtained the correct result.

Thus Newton showed that the marked irregularities of the Moon's motion fit a system controlled by universal gravitation. He did not work out the effects of the Sun's pull in complete detail; and the Moon's motion has remained a complex problem, solved with increasing detail from then till now. The ideal method would be the general one: attack the disease, not each of its many symptoms, and simply calculate the path of the Moon in the combined gravitational field of Earth and Sun. This is the "Three Body Problem": given three large masses, heaved into space with any given initial velocities, work out their motions for all time there-

142

after. Though this problem looks simple in its ingredients, it has remained a challenge for centuries; its possibilities have proved too complex for a complete explicit solution. Methods are still being explored.

7. Tides. Ocean tides had long been a puzzle, crying for some explanation connected with the Moon. Yet men found it hard to imagine a real connection and even Galileo laughed at the idea. Newton showed that tides are due to *differences* of the Moon's attraction on the water of the ocean. The Moon does not make its orbit around the Earth's center; but Moon and Earth together swing around their common center of gravity like two unequal square-dance partners. That center of gravity is 3000 miles from the center of the Earth, only 1000 miles under ground. While the Earth pulls the Moon, the Moon pulls the Earth with an equal and opposite pull (Law III), which just provides Mv^2/r to keep the whole Earth moving around the common center-of-gravity, once around in a month. The portion of ocean nearest the Moon is pulled extra hard (because it is nearer) so it rises in a lump which is a high tide. The ocean farthest from the Moon *needs* the same pull as all the rest, but is pulled less than average; so, local gravity, "g", must provide the remainder of the needed pull.[12] That makes the ocean farthest from the Moon slightly lighter; so it is pushed up into a hump away from the Earth—another high tide.

Therefore, there are two high tides in twenty-four hours. As the Earth spins, its surface travels around, while the tidal humps, held by Moon and Sun, stay still, so the tides surge up and down the shores that are carried under them. The ocean water moves around with the Earth; but the tidal humps move like a wave from shore to shore. Friction and inertia delay those surgings in a complicated manner, so high tide is not just "under the Moon" but often lags as much as ¼ of a day. The Sun also produces tides—not so big, because its greater distance makes the *differences* of pull smaller. Twice a month, when the Sun's and Moon's tides coincide, we have large "spring" tides. When the two sets of tides are out of step we have small "neap" tides.

We can estimate the "tide-generating-forces" that act on some standard chunk of material in various parts of the Earth. Take a sample chunk that weighs 30,000,000 newtons[†] at all places on the Earth's *surface.*[*] Then at Earth's *center*, E, Earth's pull on the chunk is zero; Moon's pull just provides the needed force $mv^2/$(radius EG) for the chunk's monthly motion—and calculation shows that to be 100 newtons. At all other places, A,B,C, . . . , the *needed* force is the same, 100 newtons towards the Moon.[12] But Moon's pull is only 97 at A, and 103 at C. So the *radial* pulls on a sample chunk are:

at A, 30,000,000 + 97, which will provide the needed 100 and leave 29,999,997 for effective "g"

at B, 30,000,000 + the vertical component of Moon's pull, which is now slanting. That component is ⁴⁰⁰⁰/₂₄₀₀₀₀ of 100, or about 1½. This makes 30,000,001½ for effective "g".

at C, 30,000,000 (inward) and 103 (outward), which will provide the needed 100 (outward) and leave 29,999,997 (inward) for effective "g".

Thus, at A and C the chunk is "lighter" than at B: it feels an outward tide-generating-force of 4½ newtons. That is the force that piles up the two humps, only 4½ newtons on each 3300 tons of ocean.

---

[12] The Moon travels in its orbit around the common center-of-gravity, G, in a month. And, since it always keeps the same face to the Earth, it also spins, one complete turn per month—as seen by a stationary observer, at rest among the stars. But the Earth does not keep the same face always to the Moon. The whole Earth moves around G, one revolution in a month; but it does not also have a monthly spin. Instead, it keeps a fixed orientation as it moves around G (apart from daily spin, neglected here[*]). Thus, all parts of the Earth move in circles of the same size—like the circular motion of a man's hand cleaning a window, or the frying-pan motion of Fig. 17-10b on p. 257. This motion requires every part to have the same acceleration v^2/r, where r = EG, towards the Moon.

[†] Mass of sample = 3 × 10⁷/9.8 ≈ 3 × 10⁶ kg ≈ 3300 tons. So it is a 3300-ton chunk of rock (the size of a house) or of water, or even of air.

[*] We neglect the differences of "g" from equator to poles. These differences are real, but they do not produce a noticeable effect on the oceans. That is because the oceans suffer the same effects of the Earth's spin as the "pasty" Earth of Newton's calculation (4 above); so the values of "g" are just those that will spread the oceans evenly over the bulgy spinning Earth. (I.e., we assume that daily spin has given the Earth an "equilibrium shape," and expect it to do the same for the oceans).

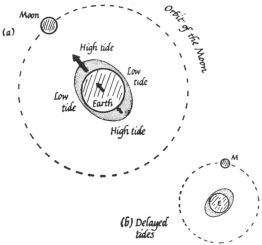

FIG. 22-23.
OCEAN TIDES ARE CAUSED BY DIFFERENCES OF MOON'S ATTRACTION
(a) The extra-large pull on ocean nearest the Moon raises one high tide. The extra-small pull on ocean farthest from the Moon lets it flow away into another high tide.
(b) Delayed tides. Actually the high-tide humps are delayed by inertia, tidal friction, and effects of rotation. As the Earth spins, they are not opposite the Moon. In most places they arrive about 1/4 cycle (6 hours) late.

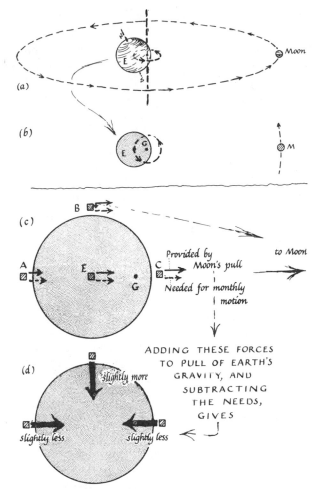

FIG. 22-24. THE TIDE-PRODUCING FORCE

8. *The Moon's Mass.* By comparing neap tides and spring tides we can separate out and compare the effects of the Sun and the Moon.[13] Newton did this and was able to make an estimate of the Moon's mass from the size of the tide it causes. In other words the Moon has always had a satellite after all, the hump of ocean which we call high tide. No independent check of the Moon's mass could be made for two centuries, until man could send up "Moon-probe" satellites.

9. *Comets.* Newton explained the nature of comets, those visitors in the solar system which had always aroused interest and even fear. (It is strange to note how comets are still regarded as mysterious things by the popular press. A tabloid

[13] At shores where there are bays and river mouths, tides may pile up to great heights; but at islands in mid-ocean the spring tides rise only about 4 ft, and the neap tides only 2 ft. Therefore: tide due to [Moon + Sun] is 4 ft, and tide due to [Moon − Sun] is 2 ft. This makes the Moon's tide 3 ft and the Sun's tide 1 ft. Thus we can estimate the ratio of Moon's to Sun's mass. However, momentum and friction make the problem very complicated.

newspaper hesitates to call an eclipse a mystery because it will be laughed at; but when a visible comet appears, or even rumors of one, most newspapers make a fuss about the "mysterious event in the heavens." This superstition survives with some of the feelings that made astrology powerful for centuries—a sad reminder of the way in which the pressures of civilized life curdled man's simple wonder into fear.)

Tycho and Kepler had shown that comets are not just "vapor in the clouds" but that they travel across the planetary orbits, making a single trip, as was thought, through the solar system. They seemed to be illuminated by sunlight, and therefore could only be seen when fairly near. Newton showed how comets move in a long elliptical orbit, with the Sun in one focus. They are controlled by gravity just like the planets; but they are small and they have far more eccentric orbits so they are only visible when they are near the Sun. Such comets travel out beyond the farthest planets, slower and slower (Kepler Law II); finally turn around the remote

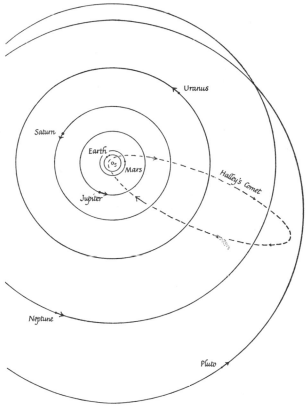

FIG. 22-25.
SKETCH OF THE SOLAR SYSTEM, WITH HALLEY'S COMET SHOWN
The most recently discovered planet, Pluto, is very small and pursues an elliptical orbit extending from within Neptune's to a much greater distance.
(Mercury and Venus are not shown)

nose of their ellipse (Kepler Law I); and after a long time (Kepler Law III), come hurtling back into our region; then they swing, at maximum speed, around the Sun and away again. The elliptical orbit can be measured and the comet's return predicted. One of the most famous, named after its discoverer Halley (who saw the *Principia* through the press) was the first case of successful prediction, 76 years from one visit to the next. Newton pointed to an earlier record by Kepler at just the right time, and predicted future returns. When these occurred on time, comets should have lost all mystery, though not their glory. Their regular return on predicted dates gives a test of our orbit-observations, and a further confirmation of the law of gravitation. We can carry the calculations back and identify some of the comets of history. For example, Newton's comet, observed by him in 1680 and expected to return in 2255, may have been the comet that was thought to herald the death of Julius Caesar.

Occasionally a comet suffers a severe gravitational perturbation in passing near a big planet, and changes to a new orbit with a different cycle of returns. That is how we know comets have small mass: *they* are affected, but the planet, pulled with an equal and opposite force, fails to show a noticeable effect.

If a comet arrives from outer space *very fast*, it swings around the Sun and away in its new direction in a hyperbola instead of an ellipse, and then it never comes back.[14]

Comets move so fast and visit so seldom that we are still waiting to apply the most modern methods of investigation to a big one. We believe they are collections of rock, dust, gas, etc. all travelling together. As they approach the Sun, they reflect more and more sunlight and look brighter and brighter. When very near they may be heated so much by sunlight that they emit some light of their own. The Sun's radiation roasts vapor out of some, adding to the light-scattering material, making these comets look bulkier and brighter. Many comets develop a "tail" of bright (but transparent) material which streams out behind and curves sideways from the orbit, *away from the Sun*. Why does this tail not keep up with the rest? The body of the comet is made up of many pieces, but these will all move around the orbit together since the Sun's gravitational pulls are proportional to the masses—remember the Myth-and-Symbol experiment. The tail however seems to be an exception. It fails to keep up with the rest and even swings sideways. This suggests some repulsive force between Sun and comet which affects the *tail proportionally more than the rest*. The tail probably contains the smallest particles—dust, perhaps, or just gas molecules. What forces affect a small particle proportionally more than a big one? Surface tension, fluid friction, and any forces that vary directly as the surface *area* of the particle; whereas masses, and therefore gravity forces, vary as the *volume*. The most likely "surface forces" on comets' dust are the pressure of sunlight and the pressure of ions streaming from the Sun. Making a particle ten times smaller in linear dimensions reduces its mass by a factor of 1000 but reduces its surface area by a factor of only 100, making surface-forces 10 times more important than they were, compared with gravitational pulls in their

---

[14] If we ask, as Newton's fellow scientists did, "Given inverse-square law attraction, what shape must a planet's —or a comet's—orbit have?", the mathematical machine replies, as it did in Newton's hands, "The orbit will be a

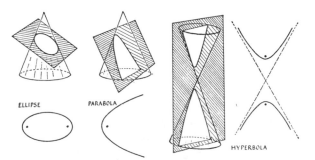

FIG. 22-26. CONIC SECTIONS

conic section with the Sun in one focus." Conic sections are curves got by taking plane slices of a solid circular cone. A cone sliced "straight across" gives a circle. If the slice is slanted, the section is an ellipse. With greater slant, just "parallel to the cone's edge," the section is a parabola. With still greater slant, the section is a hyperbola. These curves all belong to one geometrical family. Their algebraic equations are similar:

$$x^2 + y^2 = 9, \text{ a circle}$$

$$\frac{x^2}{9} + \frac{y^2}{4} = 1, \text{ an ellipse}$$

$$\frac{x^2}{9} - \frac{y^2}{4} = 1, \text{ a hyperbola}$$

The equations for parabolas look different (e.g., $x^2 = 9y$) but are closely related to the others. In physics, we meet ellipses in planets' orbits, all these curves in comets' orbits; and hyperbolas when alpha-particles are shot at other atomic nuclei. From measurements of such rebounding of alpha-particles, we can calculate the arrangement of forces that must cause these rebounds; and we find the forces must be inverse-square ones between the alpha-particle and some very tiny core of the atom it "hits." We guess that these forces are inverse square repulsions between electric charges. From further measurements we can even estimate the electric charges of different atom cores. That is how, in this century, Newtonian mechanics established the nuclear atom.

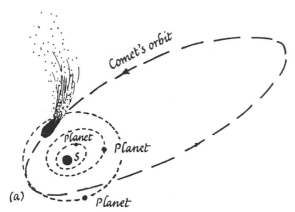

FIG. 22-27a. COMET, MOVING IN AN ELLIPTICAL ORBIT WITH THE SUN IN ONE FOCUS, PASSES THROUGH SOLAR SYSTEM

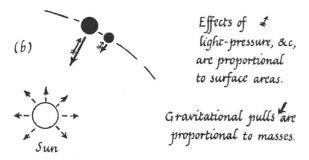

Effects of light-pressure, &c, are proportional to surface areas.

Gravitational pulls are proportional to masses.

FIG. 22-27b. FORCES ON PARTICLES IN A COMET'S TAIL
If one particle of the comet has 10 times the diameter of another, its mass, for the same density, is 10 x 10 x 10, or 1000, times as great as the other's. Gravitational forces on it will be 1000 times as big. But surface-forces, e.g., light-pressure from sunlight, will be only 10 x 10, or 100, times as great. Therefore, light-pressure matters proportionally more for small particles; less for large ones. It can push the very tiny particles in the comet's tail away.

effect on the mass. Near the Sun, the radiation is intense, also streams of protons gush out, and the pressure on the smallest particles becomes important. It is thought that such pressures drive the comet's tail away from the Sun.

10. *Gravity inside the Earth.* Newton showed by calculus that a hollow spherical shell of matter attracts a small mass *outside* it as if the shell's mass were concentrated at its center. By imagining the Earth built of concentric shells (even of different densities), Newton was able to proceed with the "apple and Moon" comparison, knowing that the Earth would attract as if its whole mass were at its center. He also showed that a hollow spherical shell would exert *no force at all* on a small mass *inside* it. This result is not much use in treating the Earth's gravity, though it is very important in the corresponding theory of electric fields. There it provides a first-class test of the inverse-square law of forces between charges. We shall derive it in the chapter on electric fields.

These two results for a spherical shell give an interesting picture of the gravitational field of a solid uniform sphere. Outside, the field decreases with an inverse-square law: "g" varies as $1/R^2$, where R is the distance from the center. If we are at an inside point, we are *inside* some of its concentric spherical shells, and we lose their attraction completely. We are *outside* the remaining central

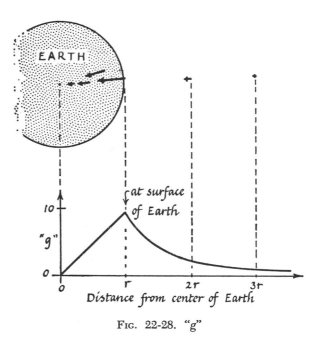

FIG. 22-28. "g"

batch of shells. These make a smaller attracting mass, but we are closer to the center. The resultant attraction makes "g" vary directly as R, inside the sphere.

11. *Artificial satellites.* Newton pointed out that any projectile is an Earth-satellite. Suppose a man on a mountain top fires a bullet horizontally. A slow bullet falls to the ground in a "parabola," which has its focus just below its nose. The path is really a Kepler ellipse, with the lower focus at the Earth's center. Parabola and ellipse are indistinguishable in the small part of the orbit observed before the ball hits the ground. (To obtain a true parabola we would need a great flat Earth, not a round one with radial directions of "g"). A faster bullet still makes an ellipse, but not so eccentric—still faster, an even rounder ellipse. One fired fast enough would go on around the Earth, like a little moon, travelling round its circular orbit again and again (provided the man got out of the way one "little-moon-month" after firing the bullet). Here was Newton's picture of an artificial satellite. It and the Moon would form a Kepler-Law-III group, with Earth as owner. We now have satellites that do this.

Fire the bullet still faster than for a circular orbit,

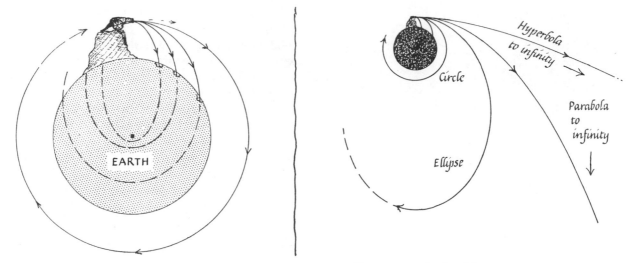

Fig. 22-29. Earth-Satellite Orbits (from Newton's sketch)
Where elliptical orbits are drawn through the Earth, they show how the original path would continue outside a tiny concentrated mass. They do not allow for the decrease of attraction inside the Earth.

and its path would be an ellipse with the Earth's center in the nearer focus this time. Still faster and the ellipse would elongate to a great parabola. Faster still and the bullet would leave the Earth in a hyperbola and would never return. The velocity necessary for such "escape" can be calculated—an important matter for future space travel, and important long ago for speedy gas molecules that escaped from the atmosphere.

12. *Planetary perturbations. The great discovery.* The major influence on a planet's motion comes from the Sun; but the other planets, acting with the same universal gravitation, also apply small forces which "perturb" the simple motion. Newton began the study of these perturbations. For example, the great planet Jupiter attracts neighboring Saturn enough to make noticeable changes in Saturn's orbit. The attraction changes in direction, since Jupiter and Saturn are both moving in their orbits; and it changes greatly in amount, as the planets move from greatest separation to closest approach.[15] This small, tilting, changing pull builds up changes of motion that accumulate in tiny perceptible changes of orbit. Newton estimated this effect and showed that it fitted with observed peculiarities of Saturn's motion. However, the general problem is very difficult, and Newton only made a start on it.

A study of planetary perturbations looks like fiddling with trivial details; yet over a century later it led to a great triumph, the discovery of a new planet. Before that the first planet beyond the five known to Copernicus was discovered by telescopic observation. In 1781 Herschel noticed a star that looked larger than its neighbors and was found to move. This proved to be a planet, soon named Uranus. The new planet was found to be twice as far away as Saturn, and its orbit radius and "year" fitted into Kepler's Law III.

The continued observations of Uranus showed small deviations from the Kepler orbit. Some of these could be explained as perturbations due to Saturn and Jupiter. However an unexplained "error" remained—a mere 1/100 of a degree in 1820. Some astronomers questioned whether the inverse-square law was exactly true of nature; others speculated about another, unknown planet perturbing Uranus. That was ingenious but posed almost impossible problems. However, two young mathematicians, Adams in England and Leverrier in France, set out to locate the planet. It is hard enough to compute the effect of one known planet on another. Here was the reverse problem, with one of the participants quite unknown: its mass, distance, direction and motion all had to be guessed and tried from the tiny residual deviations of Uranus from its Kepler orbit.

Adams started on the problem as soon as he finished his undergraduate career. Two years later he wrote to the Astronomer Royal telling him where to look for a new planet. Adams was right within 2°; but the Astronomer Royal took no notice, beyond asking Adams for more information. Then, as now, professional scientists were besieged with letters from cranky enthusiasts and had to ignore them.

Meanwhile, Leverrier was working on the problem quite independently. He examined several

[15] Their distance apart changes from the sum of orbit radii to the difference: 880 million miles + 480 million to (880 − 480) million, a change in proportion 3½ to 1. This makes the perturbing attraction increase in proportion 1 to 12.

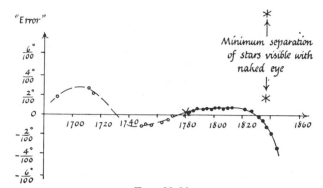

FIG. 22-30.
RESIDUAL "UNEXPLAINED" PERTURBATIONS OF URANUS
(A.D. 1650-1850)
The "error" is the difference between the observed position of Uranus and the expected position (for a Kepler orbit) after known perturbations had been subtracted. The point X marks the discovery of Uranus by Herschel. Working back to its orbit in earlier times, astronomers found that Uranus had been observed and recorded as a star in several instances. These earlier records are marked by ° on the graph.
(After O. Lodge, *Pioneers of Science*)

hypotheses, decided on an unknown planet, and finally managed to predict its position, near to Adams'. He too wrote to the Astronomer Royal, who then arranged for a careful but leisurely search. By this time other astronomers began to believe in the possibility—"We see it as Columbus saw America from the shores of Spain." Leverrier wrote also to the head of the Berlin observatory, who looked as directed, compared his observation with his new star-map, and saw the planet! The discovery raced around the world and was soon confirmed in every observatory. This new planet, discovered by pure theory, was named Neptune.

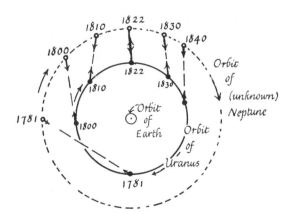

FIG. 22-31. PERTURBING FORCES ON URANUS, DUE
TO NEPTUNE
The sketch shows positions of the planets in the years marked. Before 1822 Neptune's pull made Uranus move faster along its orbit so that it reached positions ahead of expectation. After 1822 Neptune's pull retarded Uranus. (After O. Lodge, *Pioneers of Science*)

Newton's Method

Newton set forth his treatment of astronomy in the *Principia*. He was using *deduction* to derive many things from a few laws, but his treatment was essentially different from the deductive methods of the Greeks and their followers. Newton devised his theory with the help of guesses from experiment; then drew from that theory many deductions; and *then* tested as many of these deductions as he could by experiment. Thus his theory was a framework of thought and knowledge, tied to reality by experiment and clear definitions, able to make new predictions which in turn were tested by experiment. A theory, as Newton used it, "explained" a variety of mysteries by referring them to a few familiar things.

Newton's successors mistook Newton's view of gravitation. They thought he treated it as "action at a distance," a mysterious force that arrives instantaneously through a vacuum, in contrast with Descartes' picture of space filled with whirlpool vortices that transmit force and motion. Newton himself merely said that an inverse-square field of force will account for Kepler's laws and many things besides. For this "explanation" he did not need to know how the force got there. He said clearly that he did not know the *cause* of gravitation. He suggested it must be some kind of influence that travels out from every piece of matter, and penetrates every other piece, but that was only description of observed properties. He insisted he did not know its ultimate cause. *"Hypotheses non fingo"*—"I will not feign hypotheses"—he wrote fiercely at one time. He meant he would not invent unnecessary details in his description of Nature, or pretend to explanations *that could never be tested.* Yet in later writing he offered many a keen guess, at the nature of light, at the properties of atoms, and even at the mechanism of gravitation.

Newton is usually described as a genius of cold, unemotional logic and clear insight who set the style for modern science. But one of his biographers, Lord Keynes, who studied many of Newton's writings, found him a difficult recluse who treated Nature as a mystical field of magic.

"Newton was not the first of the age of reason. He was the last of the magicians, the last of the Babylonians and Sumerians, the last great mind which looked out on the visible and intellectual world with the same eyes as those who began to build our intellectual inheritance rather less than 10,000 years ago. Isaac Newton, a posthumous child born with no father on Christmas Day 1642, was the last wonder-child to whom the Magi could do sincere and appropriate homage."[16]

He felt himself the magician who had unlocked God's riddle of the solar system by reasoning from clues strewn around by God in recorded measurements, in experiments waiting to be done, in useful folklore, and even in some inspired hints in ancient writings. He succeeded in this by his extraordinary gift of concentrating continuously on a problem in reasoning, "his muscles of intuition being the strongest and most enduring with which a man has ever been gifted."[17] Could he not in the same way unlock the great riddle of Religion: discover the nature of God, explain the behavior of matter and man's mind, and reveal the whole progress of time from the original creation to the ultimate heaven? This tremendous mind aspired, as Keynes sees him, to be "Copernicus and Faustus in one."[18] Be that as it may, all the biographers, from Newton's contemporaries to Keynes and Einstein, regard Newton as the greatest mathematical scientist of a thousand years.

Newton's "Guesses"

As an architect of science Newton set a new building style, much of which is still in great use. As a thinker about science he seems to have been a remarkably lucky guesser—he guessed right more often than mere chance would make likely. He formulated laws of motion which we still use and consider very nearly "right." He guessed correctly at universal gravitation. He even made a guess, on scanty evidence, at the mass of the Earth—a guess which could not be tested at the time but had to wait for Cavendish's experiment. He argued, speculatively, that none of the solid ground can be less dense than water, or it would float up into mountains. On the contrary, the central regions must be much denser than the outer rocks. So, he guessed, the average density of the whole Earth must be between 5 and 6 times the density of water. We now know it is 5½ times! The Earth's mass is 5.5 times that of an equal globe of water. Again, Newton devised a theory of light waves that would explain both the properties of rays and the interference colors of thin films (which he discovered and measured). It was a curious scheme in which light consists of bullets accompanied by waves to arrange where they shall travel. A century later, the wave theory of light displaced and discredited the bullet

[16] "Newton the Man" by J. M. Keynes, in *Newton Tercentenary Celebrations of The Royal Society of London* (Cambridge University Press, 1947), p. 27.
[17] *Ibid.*, p. 28. [18] *Ibid.*, p. 34.

theory. For many years scientists laughed at Newton's queer mixed scheme. Now, two centuries later, we have gathered clear evidence to show that light *does* behave both as waves and as bullets, and we now hold a composite theory which has a surprising resemblance to Newton's! Once again he guessed right.

I do not think this successful guessing which characterizes Newton and other great men is due to luck or to a mysterious intuition, or divine inspiration. I believe it is due to Newton using more of his knowledge; gathering in and using chance impressions and other things barely perceived and soon quite forgotten by ordinary men; and thinking with great flexibility of mind. He had unusual intuition just because he drew upon a greater fund of detailed knowledge—he was sensitive and could remember, where most of us are insensitive and forgetful; and he was willing to turn his thinking quickly in different directions. Just as the great actor is aware of his audience and can draw upon a rich knowledge of other people's emotions and behavior, so Newton was aware of nature and could draw on a rich fund of observation. Perhaps in some respects that is where greatness lies, in sensitive awareness of the world around—the world of people or the world of things.

PROBLEMS FOR CHAPTER 12

1. In text.

★ 2. KEPLER'S LAW III

Newton guessed at universal, inverse-square-law gravitation. We express his guess in the form $F = GM_1M_2/d^2$. From his guess (a "principle") he deduced (predicted) the behavior of the Moon, planetary systems, tides, etc.

Predict Kepler's Law III by following the instructions (A) and (B) below. Suppose a Sun of mass M holds a planet of mass m in a circular orbit of radius R by gravitational attraction. Suppose the planet moves with fixed speed v, taking time T (its "year") to travel once round its orbit.

(A) *Algebra.* State in algebraic form, each of the following:
  (a) The planet's acceleration.
  (b) The force *needed* to give this planet this acceleration.
  (c) The force *provided* by gravitational attraction, if it follows Newton's law of gravitation.
  (d) The planet's velocity v in terms of R and T.

(B) *Argument:*
  (i) Now write down, as an algebraic equation, Newton's guess, that the needed force (b) is just provided by gravitational pull (c).
  (ii) Then get rid of v in this equation by using relation (d).
  (iii) Now move all the R's and T's in the equation over to the left hand side of the equation and everything else to the right hand side, thus obtaining a new (version of the old) equation.
  (iv) In the new equation, do you find R^3/T^2 on the left? (If not, check your algebra.) Do you now find the

right hand side is the same for all planets; that is, a constant which does not contain m, or R, or T?

(v) Would this new equation hold, *with the same right hand side* for other planets with different masses, orbits, and years, but with same Sun? Then does Newton's guess predict Kepler's Law III?

3. RELATIVE MASSES OF PLANETS

(a) Starting with Newton's laws of motion, and $a = v^2/R$, and universal gravitation, $F = G M_1M_2/d^2$, show how the ratio (JUPITER'S MASS)/(SUN'S MASS) can be obtained from astronomical measurements. Show your full working; do not just quote algebraic results.

(b) Make a *rough estimate** of the ratio. (See below for data)

(c) Make a similar *rough estimate** for the ratio (EARTH'S MASS)/(SUN'S MASS).

(d) From terrestrial experiments, such as Cavendish's, we can estimate the mass of the Earth. It is about 6.6×10^{21} tons. *Calculate roughly,** from (c), the Sun's actual mass, in tons.

DATA (some of which may not be needed):

Radii of planetary orbits: see table in Ch. 18.

Lengths of planetary "years": see table in Ch. 18.

Data for satellites of Jupiter: see Ch. 19. (Do not use the values of orbit-radii given in multiples of Jupiter's personal radius; but use the values in miles. The "times" are given in hours. Convert them to the units you use for all other "times" in the calculation.)

Data for Earth: personal radius $\approx$ 4,000 miles

time of revolution about axis = 24 hours

radius of orbit $\approx$ 93,000,000 miles $\approx 1.5 \times 10^{11}$ meters

1 year $\approx$ 365 days $\approx 3 \times 10^7$ seconds

Data for Moon: radius of orbit $\approx$ 240,000 miles ($\approx$ 60 Earth-radii).

Personal radius $\approx$ 1,000 miles

1 month = 27.3 days. (This is the Moon's absolute period, relative to stars.)

4. ARTIFICIAL SATELLITES

(a) Suppose an Earth-satellite pursues a circular orbit 4000 miles above the Earth's surface—that is, at radius 8000 miles from the Earth's center. From your knowledge of planetary motion, estimate the time re-

---

* In these questions where a VERY ROUGH ANSWER is asked for—to give a general idea of relative masses, or the size of some force—precise arithmetic would miss the point and give no advantage for this purpose. Therefore, you are strongly advised to proceed as follows:

(i) USE ALGEBRA UNTIL AS LATE A STAGE AS POSSIBLE.

(ii) Then insert arithmetical values, with no cancelling, and SHOW THE RESULT IN FULL FACTORS. (DO NOT CANCEL THAT FIRST "RESULT" BUT LEAVE IT UNTOUCHED, IN CASE YOU NEED TO RETURN TO IT OR READERS WISH TO CHECK IT.)

(iii) Re-copy the "result" of (ii) and make ruthless approximations to find a rough answer.

If your result is wrong by a factor of 1000 through careless cancelling, it is worthless; but if it only suffers from a 40% error due to rough arithmetic, it still carries a useful message in such problems.

quired for the satellite to make one trip around. Give your answer (i) in factors with no cancelling

(ii) reduced to a *rough numerical estimate** in minutes or hours, or days or years.

(Obtain any data you want from earlier problems above. You do not need the value of G).

(b) Nowadays we have "stationary" satellites which link one part of the Earth with another for television or for carrying telephone messages. Such a satellite receives and re-broadcasts short-wave radio signals. Therefore we want the satellite to stay in place, hovering permanently over Chicago (for example), without using any motive power once it is there.

(i) Describe the behavior of such a satellite, as seen by an observer far away from the Earth.

(ii) Calculate the height at which the satellite would have to hover. (Give your answer first in factors, then worked out *roughly** in miles.)

(c) A satellite is clocked at 90 minutes per revolution around the Earth (relative to the stars). Assuming that its orbit is circular, *estimate** its height above the surface of the Earth.

(d) Suppose a projectile could be shot out of a gun horizontally so fast that it never hits the ground but continues around the world just above the ground.

(i) How long* would it take to return to its starting point (if air resistance were negligible)?

(ii) *Estimate** its speed.

(iii) The speed asked for in (ii) is the speed that a point on the Earth's equator would have if the Earth were spinning so fast that . . . . . ?

(e) [A quick question to try on your neighbor: time limit 15 secs.] How long would a 1-ton satellite take to go around a circular orbit around the Earth, with radius a quarter of a million miles?

5. BOHR ATOM-MODEL

Bohr constructed his simplest model of a hydrogen atom with an electron pursuing a circular orbit around a massive nucleus, which exerted an electrical *inverse-square-law attraction* on it. (This form of atom-picture is now considered misleading. Yet it is still used in advertising, and even by physicists when they need a crude picture to aid rapid thinking.) The "quantum theory restriction" formulated by Bohr stated, essentially, that only those circular orbits are allowable for which

(momentum of electron) · (circumference of orbit) = $n \cdot h$

where h is a universal "quantum-constant" and n is a *whole number*: 1, or 2, or 3, 4, 5, 6, etc. . . .

Thus, $(mv) \cdot (2\pi r) = nh$ allows the atom to have its electron only in orbits, with $n = 1, 2, 3$, etc.

(a) With help from Kepler and Newton, show that the radii of allowed orbits must be proportional to n^2, so that they run in proportions 1:4:9: . . . . (Thus if an "unexcited" atom had radius x, the atom in higher excited states would have radius $4x$, $9x$, etc.

(b) For simple hydrogen atoms with $n = 1$, $r \approx 0.5$ Ångstrom Unit (0.5×10^{-10} meter), and this seems to be the "size" of the atoms. Excited hydrogen atoms in stars have been observed with n as large as 30. What would be the "size" of such atoms?

CHAPTER 13 · UNIVERSAL GRAVITATION

The idea of inverse-square-law gravitation was "in the air" when Newton made his calculations. Other scientists were speculating on a cause for Kepler's laws and asking whether planetary motions could be explained by an attraction spreading from the Sun, thinning out as it spreads. Newton rescued the question from mere speculation and extended the guess of some-kind-of-pull-from-the-Sun to universal gravitation. He tested his guess of inverse-square-law gravity by treating the Moon's motion, and by showing it led to Kepler's Laws. Further tests on Jupiter's satellites showed that the same kind of force acts between planets and satellites as between Sun and planets. So the $1/d^2$ factor in the relation $F = G M_1 M_2/d^2$ seemed well established by experimental evidence from the solar system.

The "Myth-and-Symbol" experiment guaranteed the factor M_2, the mass of the attracted body. Since all bodies fall freely with the same acceleration, g, the Earth must pull them with gravitational attractions that are proportional to their masses, $M_2 M_2'$.

Fig. 23-1. Myth-&-Symbol

Newton trusted his Law III, action equals reaction, which he considered he had tested somewhat by his pendulum experiments on momentum-conservation. Then the gravitational pull of M_1 on M_2 must be equal and opposite to the gravitational pull of M_2 on M_1. That is, $_1F_2 = {}_2F_1$. Therefore, G must be the same in the two forces below

$$_1F_2 = G(M_1 M_2)/d^2 \qquad _2F_1 = G(M_2 M_1)/d^2$$

Thus the attracted and attracting bodies are interchangeable in this story, and gravitational attraction

must also vary directly as the mass of attracting body. This seems very likely, even certain, to anyone believing in symmetry; but it can not be proved experimentally by astronomical measurements since we have no other way of estimating *astronomical* masses—until rocket explorers can bring back surveys and samples.

Trusting his general theory, Newton could estimate the ratios of celestial masses, $\dfrac{\text{Mass of Jupiter}}{\text{Mass of Sun}}$, $\dfrac{\text{Mass of Earth}}{\text{Mass of Sun}}$, and even, by guesses from tides, $\dfrac{\text{Mass of Moon}}{\text{Mass of Sun}}$; but he could not calculate actual masses separately because he did not know the value of the universal gravitation constant, G. To find the value of G required terrestrial experiments to measure the very small attraction between two *known* masses.

Measuring G

The gravitation constant, G, remained unknown for over half a century after Newton. A rough estimate of G from guesses like Newton's of the average density of the Earth showed that the attractions between small objects in a laboratory must be almost hopelessly small. The common forces of gravity seem strong; but they are due to an Earth of huge mass. And the Sun, with enormously greater mass still, controls the whole planetary system with gravitational pulls. But the pulls between man-sized objects are so small that we never notice them compared with Earth-pulls and the short-range forces between objects in "contact." It was clear that to measure G very delicate and difficult experiments would be needed.

As a desperate attempt, several scientists at the end of the 18th century tried to use a measured mountain as the attracting body. They estimated G by the pull of the mountain on a pendulum hung near it. They had to measure, *astronomically*, the tiny deflection of the pendulum from the vertical, caused by the sideways attraction of the mountain. They had to estimate, *geologically*, the mass of the mountain and its "average distance" from the pendulum. Substituting these measurements in $F = G M_1 M_2/d^2$ gave an estimate of G.

About the same time Cavendish, and later many others, measured the attraction between a large lump of metal and a small metal ball by a form of direct "weighing." Cavendish placed a pair of small metal balls on a light trapeze suspended by a long thin fiber. He brought large lead balls near the small ones in such positions that their attractions on the small balls pulled the trapeze round the fiber as axis, twisting the fiber until its Hooke's-Law forces balanced the effects of the tiny attractions.

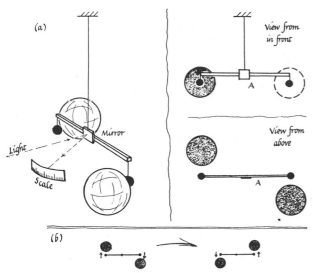

FIG. 23-2. THE CAVENDISH APPARATUS
(a) The trapeze carrying the small lead ball was hung on a very fine twistable fiber. When the big balls were brought into position, their attraction made the trapeze twist the fiber slightly. This minute twist was shown by rays of light reflected by a small mirror, A, on the trapeze.
(b) To double the measured twist, the big balls were then moved across to the "opposite" positions, so that they pulled the trapeze around the other way.

He measured the masses and the distances between small balls and large attracting balls; but, to calculate the value of G, he also needed to know the attraction forces, and to find these he needed to know the twisting spring-strength of the fiber. The fiber was far too thin and delicate for any direct measurements. So Cavendish let the trapeze and its balls twist to and fro freely with simple harmonic motion (Ch. 10) and timed the period of that isochronous motion. From that, with measurements of mass and dimensions of the trapeze, he could calculate the twisting strength of the fiber. Then, he

obtained a good estimate of G, confirmed by similar measurements made more carefully by Boys, Heyl, and others. In all cases the apparatus is so delicate that the slightest air currents will spoil the measurements. To avoid convection currents, Cavendish placed his apparatus in a box, then placed the box in a closed room and observed the apparatus with a telescope from outside the room.

Results of Measurements of G

The table shows some details of many measurements of G made over the past 220 years. It not only shows increasingly reliable values for the important quantity G, but it also gives great support to the relationships

$$\left.\begin{array}{l} F \propto M_1 \\ F \propto M_2 \\ F \propto 1/d^2 \end{array}\right\} \text{ which are combined in } F = G M_1 M_2/d^2.$$

It gives this support because it shows that a *great variety* of masses, materials, and distances all yield the same value of G, within forgivable experimental errors. If we wished to show how we know the value of G accurately (agreed for some reason or other to be universally constant) we should describe just one very good experiment; but since we want to give evidence for the validity of Newton's great theory, we give many experiments with great variety.

Modern Uses of the Cavendish Experiment

Early rough estimates of G gave a good idea of the general size of gravitational forces. The attraction between two people seated side by side is almost immeasurably small; the attraction between the Sun and the Earth is unbelievably big—a steel cable as wide as the Earth could just about replace it. And the *electric* attraction between electron and nucleus in a hydrogen atom exceeds their *gravitational* attraction by a stupendous factor of about

20000000000000000000000000000000000000.

Later measurements of G gave us the value fairly accurately, with a likely error of less than 0.2%. As recently as 1942, Heyl, at the National Bureau of Standards in Washington, made one of our most trusted measurements of this fundamental constant. Unless some new theory asked for much more precise measurements, the Cavendish Experiment would hardly be repeated. However, the apparatus has been refined into a differential gravity meter which can estimate tiny differences of gravitational field due to local deposits of rock of unusual density. This instrument is used by geologists for surveying the Earth's crust, and by oil companies to look for geological peculiarities that might yield oil. (The commercial explorers treat the instrument and its technicians with an attitude of naïve empirical hopefulness.) In one form, the two balls of a small Cavendish Apparatus are hung at *different levels*. The balls would be pulled unevenly by a shallow deposit of

Date (approx.)	Experimenter	Attracting Mass		Attracted Mass		Distance apart meters	Result G newton·m.²/kg²
		Description	Mass kg	Mass kg	Description		
A							
1740	Bouger	Mountain	*many millions of millions of kg*	*pendulum mass: a few tenths of a kg to a few kg*	pendulum	*several thousand meters*	12 × 10⁻¹¹
1774	Maskelyne	Mountain			pendulum		7 to 8 " "
1821	Carlini	Mountain			pendulum		8 " "
1854	Airy	Outer shell of Earth	3 × 10²⁰		pendulum	6,000,000 meters	5.7 " "
1854	James	Mountain	*many millions of millions of kg*		pendulum		7 " "
1880	Mendenhall	Mountain			pendulum	*several thousand meters*	6.4 " "
1887	Preston	Mountain			pendulum		6.6 " "
B							
1798	Cavendish	lead ball	167	0.8	lead ball	0.2	6.75 × 10⁻¹¹
1842	Baily	lead ball	175	0.1 to 1.5	balls of: lead, zinc, platinum, glass, brass	0.3	6.5 to 6.6 " "
1881	von Jolly	lead ball	45,000	5	metal ball	0.5	6.46 " "
1891	Poynting	lead ball	160	23	lead ball	0.3	6.70 " "
1895	Boys	lead ball	7	0.0012	gold ball	0.08	6.658 " "
1896	Braun	{ brass ball / iron ball	5 / 9	0.05	brass ball	0.08	6.66 " "
1898	Richarz and Krigar-Menzel	lead cube	100,000	1	copper ball	1.1	6.68 " "
1930 } 1942 }	Heyl and Chrzanowski	steel cylinder	66	0.05	{ platinum, glass, gold	0.1	6.673 " "

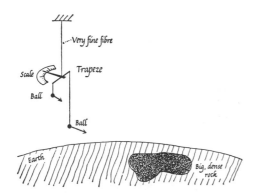

FIG. 23-3. DIFFERENTIAL GRAVITY-METER
A compact, highly sensitive form of Cavendish apparatus, with the two small balls hung at different levels. An extra dense rock nearby has a slightly greater pulling-around-effect on the lower ball than on the upper one. This is because: (i) The rock is slightly nearer the lower ball and inverse-square-law force is bigger; (ii) the pull is slightly nearer to horizontal, so that the horizontal component, which does the pulling around, is a slightly greater fraction of it. The instrument is set up, brought to an even temperature, and observed (with a telescope) in several different orientations.

dense rock nearby, and the trapeze would show a small twist when oriented suitably. This is the physicist's form of the "divining rod." To make such an instrument rugged enough to be portable and sensitive enough to be useful is a triumph of skill. Oil explorers are now replacing these differential gravity meters by more direct instruments that measure small differences of vertical g itself.

Modifications of Cavendish's experiment have been done to see whether gravitational attractions are influenced by temperature changes, or shielded by intervening slabs of matter, or dependent on crystalline form, etc. No changes have been observed—so far. G seems to be universally the same even when M_1 or M_2 includes the mass of some *releasable* nuclear energy in radioactive material: the relationship $F = G M_1 M_2/d^2$ still holds, with G the same.

Speculations

At present, most physicists regard the gravitation constant as a true constant, to be ranked with the speed of light, the charge of the electron, and a few others that seem to be the same for all matter in all circumstances, fundamental constants of the universe. However, others, bold speculators but wise ones, have suggested that G may be slowly changing as time goes on (see below). A G that was much larger in earlier times could bring gravitational forces nearer to matching electrical ones in the distant past.

Good theoretical physicists are trying to coordinate gravitational fields with electric and magnetic fields in a single "unified field theory."

Some enterprising thinkers hope to show a connection between G and other basic physical constants—possibly in connection with magnetism or perhaps in terms of the total population of atomic particles in the universe.

Time is of the Essence

Some physicists and astronomers speculating on the properties of space, time, and matter have suggested a slowly changing G, *as measured by* apparatus using *atomic clocks.* (If we used a pendulum clock to time the period of the trapeze, or if we held a pendulum near a mountain, we should find no change, because we should be comparing G with Earth's g which contains G as a factor.) This raises the whole question of TIME. What do we mean by time; and how do we know one second is the same length as the second before? We have several kinds of "clocks." Some use pendulum-swings for their constant unit of time; others use the S.H.M. of a loaded spring (Ch. 10); others use the spin of the Earth (*sidereal day*); others the Earth's revolution around the Sun (*solar year*); and others the motions of atoms (spectral lines, atomic spins, . . .); and the decay of radioactive material (Ch. 39) has been suggested. Many of these depend chiefly on *atomic* properties (e.g. radioactive decay; and even the Earth's spin, which stays practically the same even if gravity changes). But some involve gravitation directly (e.g. pendulums, the solar year). So we may have to deal with two different scales of time.

Geologists and astronomers are making good guesses at the age of the Universe from measurements of radioactivity, star temperatures, speeds and distances of nebulae. Estimates run to some 10 billion years but these are on an *atomic* scale of time. There might be such a different gravitational scale of time (e.g. by pendulum clock or by solar years) that *its* date for the beginning of the Universe stretches much farther back, perhaps even to "minus infinity."[1] That would put questions of the "beginning of the world" in a different light.

Much of this is fanciful speculation in the boundary layer of metaphysics between philosophy and science. Yet hardheaded experiments on g and G are proceeding and the next ten years may see surprising developments in our views of gravitation, with implications ranging from standards of timekeeping to theoretical cosmology.

---

[1] Here is one scheme that might be imagined. Suppose that G is slowly decreasing, as measured by "atomic" clocks. Then a pendulum clock, keeping gravitational time, must have ticked faster (judged by the spinning Earth, an atomic clock) in earlier days. So the Earth must have been spinning slower (judged by the pendulum clock) in earlier days. We shall take the pendulum clock's gravitational time as our reference standard and watch the changes in atomic time-scale. (Neither time-scale is more true than the other—only modern prejudice makes us think that atomic time is the *right* one that flows at constant rate.)

Imagine, for example, that the relationship between the two time-scales is an "exponential" one, such that, compared with pendulum clocks, atomic clocks double their rate in a trillion pendulum days. This is just one out of countless possible relationships—we choose it for a simple illustration. To make the illustration still simpler, imagine that the sliding scale connecting the two clock-schemes does not change smoothly but in sudden jumps, thus: after each trillion days by the pendulum clock, atomic clocks suddenly double their rate—twice as many atomic ticks to the pendulum day as there were before. (Such a scheme of jumps is hopelessly unlikely. Calculus offers to deal just as easily with a smooth, exponential version, which might be true of nature.) Then for each trillion days that we travel back into the past by pendulum clock, we should find atomic clocks running only half as fast—half as many atomic ticks to the pendulum day. Counting back from the present, the first trillion days would be the same on both time-scales; for the next trillion pendulum-clock days the atomic clocks would run at half rate and register only ½ trillion days; and so on. As we travel back into the past in trillion-day periods by the pendulum clock, the tally runs:

PENDULUM CLOCK: 1 trillion days + 1 trillion
+ 1 trillion + . . .

ATOMIC CLOCK: 1 trillion days + ½ trillion
+ ¼ trillion + . . .

The second series never adds to more than 2 trillion, but the first mounts to infinity. So while we count back to a definite beginning of time by atomic clocks (2 trillion days ago in this example), the pendulum clock's tally runs back from now to minus infinity.

In this example, if we measure time by radioactive changes or by the Earth's spin (which defines the sidereal day), we should find the Universe beginning some 2 trillion days ago, but we could never get back to that beginning by counting pendulum days or solar years.

PROBLEMS FOR CHAPTER 13

★ 1. Newton guessed at universal, inverse-square-law gravitation. We express his guess in the form $F = G M_1 M_2 / d^2$. From his guess he deduced (predicted) the behavior of the Moon, planetary systems, tides, etc.

(a) Say what each letter in the relation above stands for and give its proper units in the meter, kilogram, second system. Copy the example, then proceed similarly for the rest of the letters. (*Example*: "G is a universal *constant*, the same for all attracting bodies. It is measured in newtons • meters^2/kg^2.")

(b) G is a universal constant measured by Cavendish and others. If we use newtons for the force, and kg for the masses, and meters for the distance, G has the value 6.66×10^{-11} (or 0.0000000000666) newtons • meters^2/ kg^2. From this, make a rough estimate of the mass of the Earth as follows:

   (i) Using the relation above, with the value of G above, calculate the attraction of the Earth on a 0.40-kg apple near the surface. (Assume the attraction is the same as if Earth's mass were all at its center, 4000 miles from apple.) The radius of the Earth is about 4000 miles or about 6,400,000 meters. Call the mass of the Earth M kg.*

   (ii) Using your ordinary knowledge of physics, calculate the *weight* of (= pull of the Earth on) the 0.40-kg apple in *newtons*.

   (iii) Assuming the answers to (i) and (ii) are the same, write an equation and solve it for the mass of the Earth. This will be in kilograms. Convert it to pounds, then tons. (1 kg ≈ 2.2 pounds)

   * Since in stage (i) you do not know the mass of the Earth, you must call it M in your answer to (i).

2. HOW BIG IS GRAVITATIONAL ATTRACTION?

As an indication of the size of the Sun's pull on the Earth, carry out the following calculations roughly. Suppose the Sun's gravitational attraction could be replaced by a steel wire running from the Sun to the Earth, the wire's tension holding the Earth in its orbit. Good steel has a breaking stress of 100 tons-weight per square inch.

(a) Calculate *very roughly* the cross-section area of the wire which could just hold the Earth in its orbit.
(b) Calculate *very roughly* the wire's diameter.

Data: $G = 6.7 \times 10^{-11}$ newtons • meters^2/kg^2
        ≈ 1.0×10^{-9} poundals • feet^2/pounds^2
     Distance of the Sun from Earth is 93 million miles.
     Mass of the Sun: about 2×10^{27} tons
     Mass of the Earth: 6.6×10^{21} tons.

3. HOW SMALL IS GRAVITATIONAL ATTRACTION?

Calculate roughly the gravitational attraction between two boys assumed spherical, one a 70-kilogram boy, the other a 90-kilogram boy, seated with their centers 0.80 meter apart.

4. OTHER FORCES: COMETS

Comets are probably collections of separate solid lumps, dust, and gas.
(a) Explain why, if the motion is controlled by gravity, you would expect to find the comet travelling as a whole without changing shape (big lumps and little lumps keeping pace together), and not having a tail that lags behind or swings away.
(b) What type of forces would be needed to "explain" the behavior of comets' tails? (State the *essential characteristic* of these forces by describing their mathematical form rather than their physical nature.)
(c) Give clear reason for your answer to (b).

A COMMENT ON NEWTON'S PROOF OF
KEPLER'S LAW I—ELLIPTICAL ORBITS

Here is a sketch to remind you that Newton's proof (in the "black box" below) is only a piece of grinding-round of mathematical machinery, and not in itself a new piece of scientific knowledge.

INPUT $\longrightarrow$ "BLACK BOX" $\longrightarrow$ OUTPUT

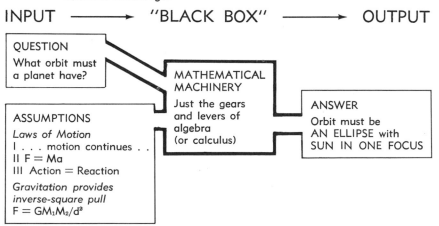

QUESTION
What orbit must a planet have?

ASSUMPTIONS
Laws of Motion
I . . . motion continues . .
II $F = Ma$
III Action = Reaction
Gravitation provides inverse-square pull
$F = GM_1M_2/d^2$

MATHEMATICAL MACHINERY
Just the gears and levers of algebra (or calculus)

ANSWER
Orbit must be AN ELLIPSE with SUN IN ONE FOCUS

CHAPTER 14 · SCIENTIFIC THEORIES
AND SCIENTIFIC METHODS

~~~~~~

"A time to look back on the way we have come, and forward to the summit whither our way lies."

—J. H. Badley

The Fable of the Plogglies

"The prescientific picture is represented by a little story about the 'plogglies.'

"According to this story, there were once two very perplexing mysteries, over which the wisest men in the land had beat their heads and stroked their beards for years and years. . . . Whenever anyone wanted to find a lead pencil he couldn't, and whenever anyone wanted to sharpen a lead pencil the sharpener was sure to be filled with pencil shavings.

"It was a most annoying state of affairs, and after sufficient public agitation a committee of distinguished philosophers was appointed by the government to carry out a searching investigation and, above all, to concoct a suitable explanation of the outrage. . . . Their deliberations were carried out under very trying conditions, for the public, impatient and distraught, was clamoring ever more loudly for results. Finally, after what seemed to everyone to be a very long time, the committee appeared before the Chief of State to deliver a truly brilliant explanation of the twin mysteries.

"It was quite simple, after all. Beneath the ground, so the theory went, live a great number of little people. They are called plogglies. At night, when people are asleep, the plogglies come into their houses. They scurry around and gather up all the lead pencils, and then they scamper over to the pencil sharpener and grind them all up. And then they go back into the ground.

"The great national unrest subsided. Obviously, this was a brilliant theory. With one stroke it accounted for both mysteries."

—Wendell Johnson[1]

~~~~~~

By the eighteenth century, science was being moulded into a body of experimental knowledge connected by logical thought. Look at the ingredients of this system that we call science.

Why do we call Newton's work good theory and the plogglies a bad theory? What makes good theory good; and why do we put such trust in theory today? We shall spend this chapter discussing such questions.

We shall not give a compact definition of scientific theory or scientific knowledge—that would make a mockery of their varied nature and great importance. You need to develop an educated taste for them as you do for good cooking; in a sense, scientific theory is a form of intellectual cookery. All we can do here is provide some general background and vocabulary for your own thinking. This chapter

[1] From *People in Quandaries* (Harper and Brothers, New York, 1946).

offers comments on the kind of theory developed by Newton's day. In other books you will find changes in the ways in which scientists construct and use theories and value them—and trust them. You would be well advised to read the present chapter quickly, to watch its to-and-fro discussion rather than to extract any final answers. Your view of science must be of your own making.

VOCABULARY

Here is a dictionary of terms for use in your thinking:

Facts

Most physical scientists believe they are dealing with a real external world—or at least they act as if they believe that in building up their first scheme of knowledge. Even if they have philosophic doubts, they start with "sense impressions" or "pointer-readings on instruments" as their *facts* of nature.

We trust such facts because they are agreed on by different, independent observers. In common life, our facts may be vague—e.g., "Uncle George is bad tempered." In physical science they are usually definite measurements, the results of experiment—e.g.:

the crystal has 8 faces,
this sheet of paper is 8.5 inches wide,
aluminum is 2.7 times as dense as water,
a freely falling stone gains 32 ft/sec in each second,
orbit of Mars is twice as wide as orbit of Venus,
the gravitation constant is 6.6×10^{-11} MKS units,
an atom is a few times 10^{-10} metres wide.

To be completely clear and true, each of these needs some commentary: definitions of terms, explanation of accuracy, limitations of applicability; but among scientific colleagues we usually leave these unsaid—just as a family may agree that Uncle George is bad tempered without worrying over accurate definitions of temper. As the list progresses we get farther and farther from direct sense-impressions, and our "facts" are more and more dependent on our choice of theory. When we get to the statement "the diameter of a hydrogen atom is 10^{-10} meter," the "fact" has little meaning unless we say what behavior of atoms we are dealing with and even what theory of atoms we are using to express the behavior.[2]

Nevertheless, we do have a vast supply of facts that we trust as coming more or less directly from experiment. They have the essential quality of the *Uniformity of Nature*; they are the same on different days of the week and in different laboratories, and they are the same for different observers. "Are your results repeatable?" is one of the first questions the research director asks the young enthusiast.

Laws

We try to organize facts into groups and extract common pieces of behavior (e.g., all metals carry electric currents easily; stretch of a spring varies as load). We call the extracted statement or relation a rule, a *law*, occasionally a principle. Thus a law is

a generalized record of nature, not a command that compels nature. Some scientists go further and idealize laws. They take each law to be simple and exact; but then they gather a guide-book full of real knowledge telling them just how nature fits the ideal law and within what range, etc. This invisible guide-book is what distinguishes the experienced scientist from the amateur who only knows the formal wording of the laws. It is no handbook of densities and log tables, but a very valuable pocket book of understanding—theory and experimental knowledge combined.

When we are trying to extract a law, we usually restrict our attention to particular aspects of nature. When we are finding Hooke's Law, our spring may be twisting, the loads may be painted different colors, the loads may even be evaporating; but we ignore those distractions. Or our spring may be growing hotter in an overheated laboratory; and then we find the stretch changing less simply. Discovering that temperature affects our measurements we arrange to keep the temperature constant. (This is an important precaution when we investigate gas expansion. Rough experiments suggest it is negligible for investigations with steel springs, but careful measurements show that good spring balances should be "compensated for temperature.")

Most laws in physics state the relationship between measurements of two quantities. For example:

PRESSURE $\propto 1/$VOLUME
FORCE $\propto M_1M_2/d^2$

Almost all laws can be reworded with the word "constant" as their essential characteristic:

TOTAL MOMENTUM remains constant in any collision

PRESSURE $\cdot$ VOLUME $=$ constant
$F \cdot d^2/M_1M_2 =$ constant

We look for laws because we enjoy codifying these regularities in the behavior of nature.

Concepts

In ordinary discussion, "concept" is a highbrow word for an idea or a general notion. In discussing science we shall give it several meanings.[3]

A. Minor Concepts

(i) *Mathematical Concepts* are useful tool-ideas, such as: the idea of direct proportionality or variation (e.g. STRETCH $\propto$ LOAD); the idea of a limit (e.g. pressure *at a point*; speed as *limit* of $\Delta s/\Delta t$).

---

[2] Measurements of an atom's size by collisions will yield different results if the collisions are made more violent, as atoms then squash to smaller size. If the bombardment is still more violent, the estimate may be much different. And, in all cases we have to make some theoretical assumptions —which are necessarily indirect.

Consider the fable of the demon tailor. Wishing to estimate the diameter of a man, he first runs a tape-measure around the man's waist, pulls the tape away, reads the man's circumference on it and divides by 3 (since 3 is near enough to 3.14 for such a rough guess). But then, with gruesome enthusiasm, he uses a thin steel wire instead, pulls it taut around the man's abdomen, cutting him as in cutting cheese. The wire ends encircling the man's spine, and the result is a diameter of 3 inches or less.

[3] This follows an excellent discussion by James B. Conant in a report on "The Growth of the Experimental Sciences" (Harvard, 1949). For more detailed discussion of the "tactics and strategy of science," see the same author's *On Understanding Science*.

(ii) *Name Concepts*: the ideas in some descriptive names that help us to classify and discuss. We may name a group of materials (e.g. *metals*) or a common property (e.g. *elasticity*).

(iii) *Definition Concepts*: the ideas that we invent and define for our own laboratory use. These may be manufactured from simple measurements (e.g. pressure from force and area; resultant of a set of forces; acceleration $= \Delta v/\Delta t$). Or they may describe some arrangement (e.g. constant temperature; equilibrium of a set of forces).

B. *Major Concepts*

(iv) *Scientific Concepts*: useful ideas developed from experiment such as:

the resultant of a set of vectors treated as things that add geometrically

heat as something that makes things hotter

momentum as a useful quantity to keep track of in collisions

a molecule as a basic particle.

(v) *Conceptual Schemes*: more general scientific ideas that act as cores of thinking, such as:

heat as a form of molecular motion

heat as a form of energy

the Copernican picture of the solar system

Newton's Laws of Motion

the picture of the atmosphere as an ocean of air surrounding the Earth

(vi) *Grand Conceptual Schemes* such as:

The whole Greek system for planetary motions etc.

Newtonian gravitational theory

Conservation of Energy

Conservation of Momentum

Kinetic Theory of Gases

Speculative Ideas

Most scientific concepts arise from experiment, or are vouched for by experiment to some extent. Other parts of scientific thinking are pure speculation, yet they may be helpful, and they are safe as long as we remember their status. We might label these *speculative ideas*. The crystal spheres were a speculative idea—invisible and quite uncheckable. In fact, the Ptolemaic scheme was not ruined when a comet was found to pass through the spheres: only the spheres were smashed. In examining any conceptual scheme, be careful to sort out its necessary concepts from the speculative ideas that accompanied its birth.

"Theory" and "Hypothesis"

Many scientists would call a grand conceptual scheme a *theory* and they would say that a speculative idea is much the same as a *hypothesis*. Both these words have become vague in general use, and almost confused with each other; so it might be better to avoid using them. However, we shall use them here; and you may profitably use them, distinguishing them as follows:

Hypotheses are single tentative guesses—good hunches—assumed for use in devising theory or planning experiment, intended to be given a direct experimental test when possible.

Theories are schemes of thought with assumptions chosen to fit experimental knowledge, containing the speculative ideas and general treatment that make them grand conceptual schemes.

THE BUILDING OF SCIENTIFIC KNOWLEDGE

Our knowledge of nature is first gained by *induction*, by extracting general rules from experimental data. Then, when we trust our rule we assume that nature will do the same thing next time—we bet on the Uniformity of Nature. If you look back on early astronomy (and on your own early laboratory work) you will see that although inductive knowledge is reasonably sure—e.g. planetary paths, Hooke's Law—it is not very fruitful in explanations or predictions. For greater, more fruitful knowledge we turn to *deductive theory*. There we start with assumptions and rules—guessed at, snatched from experiment, modelled by analogy, or invented as speculative ideas—and we make predictions and explanations; we build a sense of knowledge. But, to avoid the mistakes of the ancient philosophers, we must certainly test the predictions that emerge. We should also ask where the rules for the theory's starting point come from.

All through your study of science you should watch for the assumptions that are built into theories and check on their wisdom. Too many assumptions may lead to too much magic. "Words and magic were in the beginning one and the same thing." (Sigmund Freud, *First Lecture*.)

Look back on a simple question about demons.

Demons

Start a ball rolling along a table. How do you know that it is friction that brings a rolling ball to

158

a stop and not demons? Suppose you answer this, while a neighbor, Faustus, argues for demons. The discussion might run thus:

You. I don't believe in demons.
Faustus. I do.
You. Anyway, I don't see how demons can make friction.
Faustus. They just stand in front of things and push to stop them from moving.
You. I can't see any demons even on the roughest table.
Faustus. They are too small, also transparent.
Y. But there is more friction on rough surfaces.
F. More demons.
Y. Oil helps.
F. Oil drowns demons.
Y. If I polish the table, there is less friction and the ball rolls farther.
F. You are wiping the demons off; there are fewer to push.
Y. A heavier ball experiences more friction.
F. More demons push it; and it crushes their bones more.
Y. If I put a rough brick on the table I can push against friction with more and more force, up to a limit, and the block stays still, with friction just balancing my push.
F. Of course, the demons push just hard enough to stop you moving the brick; but there is a limit to their strength beyond which they collapse.
Y. But when I push hard enough and get the brick moving there is friction that drags the brick as it moves along.
F. Yes, once they have collapsed the demons are crushed by the brick. It is their crackling bones that oppose the sliding.[4]
Y. I cannot feel them.
F. Rub your finger along the table.
Y. Friction follows definite laws. For example, experiment shows that a brick sliding along the table is dragged by friction with a force independent of velocity.
F. Of course, same number of demons to crush, however fast you run over them.
Y. If I slide a brick along the table again and again, the friction is the same each time. Demons would be crushed in the first trip.

F. Yes, but they multiply incredibly fast.
Y. There are other laws of friction: for example, the drag is proportional to the pressure holding the surfaces together.
F. The demons live in the pores of the surface: more pressure makes more of them rush out to push and be crushed. Demons act in just the right way to push and drag with the forces you find in your experiments.

By this time, Faustus' game is clear. Whatever properties you ascribe to friction he will claim, in some form, for demons. At first his demons appear arbitrary and unreliable; but when you produce regular laws of friction he produces a regular sociology of demons. At that point there is a deadlock, with demons and friction serving as alternative names for a set of properties—and each debater is back to his first remark.

You realize that friction has only served you as a name: it has established no link with other properties of matter. Then, as a modern scientist, you start speculating on the molecular or atomic cause of friction, and experimenting to test your ideas. Solids are strong; they hang together. Their component atoms must attract with large forces at short distances. When solid surfaces slide or roll on each other, small humps on one get within the range of atomic attractions of local humps on the other and they drag each other when the motion tries to separate them. Friction, then, may be an atomic dragging, which is likely to make one surface drag small pieces off the other. That has been investigated experimentally. After a copper block has been dragged along a smooth steel table, microphotographs show tiny copper whiskers torn off on to the steel. Also chemical tests show that a little of each metal rubs off on to the other.[5]

At last you have a good case for *friction*: it is a scientific name for some well-ordered behavior that we can now link with other knowledge. It is atomic or molecular dragging, caused by the same forces that make wires strong and raindrops round. Its mechanism can be demonstrated by photographs and by chemical analysis. Its laws can even be predicted by applying our knowledge of elasticity to the small irregularities of surfaces. Friction has joined other phenomena in a general explanation.

---

[4] If Faustus has the equipment he should offer you a microphone attached to a glass table, with connections to an amplifier and loudspeaker. Then if you roll a steel ball along the table you will indeed hear noises like crushing demons.

[5] We can even show that when a *copper* block rubs on another block of *copper*, tiny pieces of copper—invisibly small—are exchanged between the two blocks. No chemical analysis could tell one block's copper from the other's; so this interchange seems impossible to detect. Yet it is now easily shown—and measured—by labelling one block with radioactive copper.

And now we can state the full case against demons: they are arbitrary, unreasonable, multitudinous, and over-dressed. We need to imagine demons with special properties to explain each phenomenon in turn: therefore we need many kinds, each with behaviors chosen to fit the facts. We now prefer something more economical and comfortable; a consistent body of knowledge, with strong ties to experiment—and with cross checks and interlinkages to assure us of validity—all expressed in as few general laws as possible. Even where we meet new events that we cannot explain, we would rather speculate cautiously than invent a demon to calm some fear of mystery.

Good Theory

Now we can return to the contrast between plogglies and Universal Gravitation. The plogglies were specialized demons. The author of the fable, a psychologist who offers it for therapeutic purposes, discusses it as an example of prescientific or magical theories that explain the working of nature by unpredictable gods or demons. He states his overall objection to the theory: "The only thing wrong with it was that there aren't any plogglies." There many modern physicists would disagree. They would not mind the plogglies being a fiction (like any "model" in science) but they would call the plogglies bad theory because they are too expensive. The plogglies were invented and endowed with two special behaviors to explain two sets of events, and they do not explain anything else. They are an "ad hoc" theory, a theory concocted just "for this purpose." There is nothing wicked about ad hoc theories— they may even turn out to be true—but they are weak, usually little more than narrow hypotheses loaded with faith. Labelled merely "ad hoc assumptions" they may be useful signposts for honest speculation. And when they lead to explanations of other observed behavior we think better of them and may promote them to a respectable title.

Then, as theory grows from a single speculative guess to a general form of knowledge that fits many observed effects we trust it more and more. We are so pleased with its consistency and fruitfulness that we say, "It can't be wrong." Look at Newton's gravitational theory as an example of such a grand conceptual scheme. Newton started with a number of assumptions: vector properties of forces and motions; the behavior summed up in his Laws of Motion; gravitational pulls proportional to inertial

masses; inverse-square law; Euclidean geometry. Some of these were extracted from experiment; others were little more than definitions (Law I defining "zero force") and rules of procedure (Law III). But whatever their origin they were stated as starting points for deductive theory. Then step by step with clear reasoning he drew his "explanation" of the solar system from them. We call this good theory because it is economical. Starting from general assumptions, Newton tied together in a single scheme many things that had seemed disconnected:

Moon's circular motion
disturbances of Moon's simple
motion
planetary motions
(Kepler's Laws I, II, III)
planetary perturbations
motion of comets
tides
bulge of the Earth
differences of gravity
precession of equinoxes

ALL
RELATED
by inverse-
square-law
gravitation
and a
spinning
Earth

Deductive Theory and Scientific Knowledge.

The fanciful picture of Fig. 24-1 shows some of the construction of good theory. Inductive gathering of knowledge must come first. Then, when the time is ripe, the theory may be brewed from a complex mixture of ingredients. At early stages some of the assumptions should be drawn off into preliminary tests (as in Newton's test of inverse-square gravity with the Moon). At a later stage the predictions yielded by the theory should be put to experimental tests—as retrospective checks on the original assumptions. We judge a theory not by how "right" it is but by how helpful, how it suggests experiments or promotes thought. To many a scientist, however, the full value of a great theory is not just in fruitful predictions but in a deep sense that it gives of sureness of knowledge.

In a way the picture shows the making of scientific knowledge rather than just of theory. It is obviously a complex method that will take many forms.

"THE Scientific Method"

Now you can see why we say there is not one single scientific method but many. Francis Bacon (~ 1600) advocated a formal scientific procedure:

make observations and record the facts
perform many experiments and tabulate the results
extract rules and laws by induction

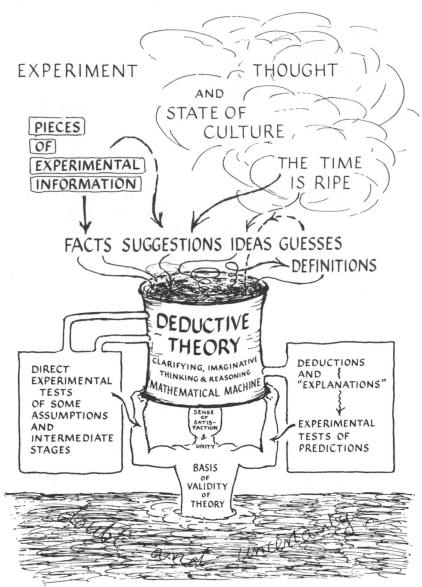

EXPERIMENT   THOUGHT

AND
STATE OF
CULTURE

THE TIME
IS RIPE

PIECES
OF
EXPERIMENTAL
INFORMATION

FACTS SUGGESTIONS IDEAS GUESSES
DEFINITIONS

DEDUCTIVE
THEORY
CLARIFYING, IMAGINATIVE
thinking & reasoning
MATHEMATICAL MACHINE

SENSE
OF
SATIS-
FACTION
&
UNITY

DIRECT
EXPERIMENTAL
TESTS
OF SOME
ASSUMPTIONS
AND
INTERMEDIATE
STAGES

DEDUCTIONS
AND
"EXPLANATIONS"

EXPERIMENTAL
TESTS OF
PREDICTIONS

BASIS
OF
VALIDITY
OF
THEORY

FIG. 24-1.

and this earnest beginning was elaborated into *THE Scientific Method*, advocated by some logical enthusiasts even to this day:[6]

> make observations and extract rules or laws
> formulate a tentative hypothesis (a guess, which may be purely speculative)
> deduce the consequences of the hypothesis combined with known laws
> devise experiments to test those consequences.

IF EXPERIMENT CONFIRMS	IF EXPERIMENT REFUTES
hypothesis, adopt the hypothesis as true law; and then proceed to frame and test more hypotheses.	hypothesis, look for an alternative hypothesis.

Real scientific enquiry is not so "scientifically logi-cal" or so simple. (We do follow some such scheme unconsciously, as James B. Conant points out, when we look for a fault in our car lights, or when we deal with water dripping from the ceiling of our apartment—but we rightly say that our quick process of guessing and looking is plain common sense.)

Scientific Methods

In the real development of science we approach our problems and build our knowledge by many

---

[6] The same "scientific approach" is advocated by some experts in other fields—e.g. the social sciences. There it offers useful guidance and criticism, but if enforced with blind enthusiasm it will probably restrict progress just as it would in physics. Besides, how do we know that knowledge in other fields can be made to crystallize in the form that fits physics?

methods: sometimes we start by guessing freely; sometimes we build a model for mathematical investigation, and then make experimental tests; sometimes we just gather experimental information, with an eye open for the unexpected; sometimes we plan and perform one great experiment and obtain an important result directly or by statistical sorting of a wealth of measurements. Sometimes a progressive series of experiments carries us from stage to stage of knowledge—the results of each experiment guiding both our reasoning and our planning of the next experiment. Sometimes we carry out a grand analysis thinking from stage to stage with a gorgeous mixture of information, rules, guesses, and logic, with only an occasional experimental test. Yet experiment is the ultimate touchstone throughout good science, whether it comes at the beginning as a gathering of empirical facts or at the end in the final tests of a grand conceptual scheme.

How far scientists' theoretical thinking will develop at a given time depends on the state of knowledge and interest—on whether the time is ripe. When the general climate of opinion is ready for a change of outlook or a new idea, a scientific suggestion may take root where it might have starved a century before; and this control of the advance of understanding by the intellectual and social climate is still true today.

When the time is ripe, the same problem is often attacked by many scientists simultaneously and the same solution may be discovered by several. Yet one scientist may get the credit for reaping the harvest—quite rightly if he is the only man with enough insight or skill to carry the innovation through. In Newton's day new interest in motion, general thinking about the planets, Kepler's discoveries, new studies of magnetism and its forces, ... and new attitudes towards experimenting and scientific knowledge, all made contributions of facts and attitudes and interests—the time was ripe for the great development. Hooke, Wren, Halley, Huygens, and others were all jumping to reach a unified theory for celestial and earthly motions. Each succeeded in grasping some parts of the solution, but it was Newton who gave the complete solution in one grand scheme, making "not a leap but a flight."

Scientific Method: Sense of Certainty

Above all, most scientific knowledge—facts, concepts, conceptual schemes—is built up by a crisscross process of investigation and reasoning from several angles. We do not push straight ahead along one narrow path of brilliant discovery—like a romantic story of making better toothpaste—but we investigate nature first along one line, then along another; and then we make still another guess and test it; and so on. As time goes on, we gather our new concepts from several lines and check them from different viewpoints; and our strong sense of good knowledge is assured by the agreement from different lines of inquiry.

The modern physics of atoms and nuclei is a particularly good example. That region of study is like a great central room with, say, seven closed doors around it. Scientists looked in at one door, and got a glimpse of micro-nature and its mysterious ways. Then through another door: quite a different view. Then through another—and then a comparing of notes. (For example, radioactivity gave one view; electron streams quite another; and the photoelectric effect raised new problems. X-rays gave still another view; and presently some knowledge of X-rays linked up with radioactivity and some with the photoelectric effect, and in still another way X-rays confirmed earlier measurements of atomic sizes.) Finally by checks and comparisons between different views a consistent scheme emerged, a *picture* was formed, to describe micro-nature. We describe that picture with everyday words from the large "macro" world around us (atoms are *round*, electrons are *small*, X-rays *travel like light*). So it is not a true description—whatever true may mean—but a "model" to enable us to describe our experience of the micro-world in familiar terms. This scheme, our model and its rules—our descriptive theory —is still being modified and extended. If we discover new experimental facts that fit, we enjoy the confirmation. If new facts conflict with predictions from our model, we modify the model—changing as little as possible, from natural conservatism. And if we discover new facts that go beyond the scope of our model we extend it. (When we found that atoms are easily pierced by fast alpha-particles, we stopped saying they are solid round balls and described them as hollow round balls.)

Our knowledge has grown already to be a comprehensive conceptual scheme that we trust because it fits with our views through many doors. Though we shall make many changes, though we may change our whole scheme of thinking about atomic physics, we already hold much knowledge with a confidence that comes from its consistency with every experimental aspect. To the outside critic seeing us looking through a single door, the evidence seems frail and the wealth of speculation too great. But those who are building the knowledge say, "We are sure we are on sound lines because if we were seriously wrong some inconsistency would show up somewhere, some clash with at least one of our experimental viewpoints—trouble will out."[7]

Building this sense of assurance is the essence of scientific methods. Professor Ernest Nagel has suggested that if there is a single scientific method it lies in the way in which scientists check and countercheck their knowledge by experiment and reasoning from several angles, so that they feel that their knowledge is warranted, that its validity is assured.

THE UNFINISHED PRODUCT— UNDERSTANDING

Models

That is why science seems complicated to learn, and even difficult to trust at first: we gain our knowledge by repeated attacks from different angles and we base our belief on the consistency of that knowledge. We do not necessarily believe that the picture of nature we thus form *is* the real world. Many scientists say it is simply a *model* that works.

It is easy to see that our picture of atomic structure is only a model—the invisible atom described in terms of large visible bullets and baseballs and large forces that we can feel like weights and the attraction of magnets. Yet it is uncomfortable to realize that we do not know what an atom is "really like," and can only say that it "behaves as if. . . ." And yet, with the progress of invention, microscope . . . electron microscope . . . ion microscope

. . . , you may decide that we can see real atoms and not just a model of them—there is a *photograph* of tungsten atoms at the end of this book. Yet all such "seeing" of the micro-world, however clear its results, is quite indirect: the images we obtain must be interpreted in terms of the models that guided our use of apparatus. In casual talk we gladly say, "Now we know what the atoms are really like, how they are arranged and how they move about"; but in serious discussion, most scientists say, "We have only shown that our model serves well, and we have obtained some measurements of parts of our model." In a way, we use models in almost all our scientific thinking: atoms, molecules, gravity, magnetic fields, perfect springs, . . . Our models are what we use to replace plogglies—economically.

Since theory is largely a reasoned model structure, based on some facts, we can always make changes in it. Popular writers describe scientists as gaily throwing away a theory when new discoveries conflict with it; but in fact most scientists cling to their old theory desperately. When scientists do change their theory to fit increasing knowledge, it is more often by progressive modification than by a revolution.

"Crucial Experiments"

Sometimes rival theories lead to different predictions so that a "crucial experiment"[8] can decide be-

---

[7] This is like our assurance of finances. It is easy to see whether the accounts of a small store are correct, but the financial statements of a big corporation are too complex for an amateur to analyze. Yet we are confident that any major financial wrongdoing, however well concealed, would show up in the course of time. After many years of watching a company's accounts without understanding them, most shareholders maintain they are sure the company is sound. We have long had that feeling of warranty for Newton's gravitational theory, and for some general principles such as the Conservation of Energy.

[8] A fine example of a crucial experiment occurred in the history of light. Two hundred years ago there were two views of the nature of light: (A) the bullet theory advocated by Newton, and (B) the wave theory developed by Hooke and Huygens. Both accounted for the general behavior of light-rays, such as reflection and refraction, but refraction also offered a crucial test.

When a slanting beam of light rays hits the surface of a pool of water, the rays are bent to a steeper slant as they enter the water. This bending of light at a boundary is called *refraction* and it has been well known as a property of light for thousands of years. Ptolemy gave an approximate law for the amount of bending, and Willebrod Snell discovered the exact law half a century before Newton wrote on optics. Both the theories of light accounted for refraction and both predicted the exact form of Snell's Law:

(A) The bullets of light must be attracted by water as they approach it (rather like a vapor molecule returning to the liquid surface). So their momentum is changed thus:
   (i) vertical component of momentum is increased (by the action of the attractive force);
   (ii) horizontal component remains unchanged (symmetry). The resultant momentum therefore runs steeper in the water, showing the refraction of the stream of bullets. The geometry predicts Snell's experimental law. With this change of momentum, the bullets must travel *faster* in water than in air.

(B) On wave theory, the advancing lines of crests must be delayed when they meet the water, so that they are slewed around and travel on through the water in a steeper direction. This requires light-waves to travel *slower* in water than in air.

A comparison of speeds of light in water and air would be a crucial experiment to decide between the two theories. It was not until 1850—a century and a half after Newton, Hooke, and Huygens—that Foucault made the crucial experiment. He showed that light moves *slower* in water than in air. That settled the case against bullets—but only against that particular model: bullets that have constant mass and move with increased velocity, momentum, and energy in water. Choose instead bullets that have the same energy in air and water but change their mass on entering water, and you can concoct a theory that predicts Snell's Law and makes the bullets move slower in water. In this case, the escape from failure is easy though the product proves to be unruly; but almost any theory can survive the condemnation of a "crucial" test, at a cost of complicated improvements.

tween them. Even then there is no absolute decision: the defeated theory can usually be pushed and twisted into a form that survives the test—just as demons could always be endowed with extra properties. For example, Newton's demonstration of a guinea and a feather falling freely in a vacuum decides between two theories of ordinary fall:

(A) "All bodies fall with the same acceleration, but for air resistance."
(B) "Bodies have natural downward motions proportional to their weights."

Yet (B) can be polished up into agreement with the experiment by blaming the vacuum pump instead of thanking it: (B) "Bodies ... weights; but vacuum also exerts downward forces in inverse proportion." (When a barometer is demonstrated, the story will have to grow more fantastic still.)

Only in a few great cases does the decision seem final: as in the choice, for the form in which light carries energy, between smooth waves and compact bullets (quanta). There, the photo-electric effect decides resolutely in favor of quanta; or, again, when experiments with the speed of light support special Relativity, the decision seems certain. Yet even in those great cases it is the weight of several lines of evidence that decides rather than an inescapable proof by a single experiment.

Intellectual Satisfaction

Thus the test of good theory is not success *vs.* failure but remains simplicity and economy *vs.* increasing complexity or clumsiness. The best theory is the one that is most fruitful, economical, comprehensive, and intellectually satisfying.

We expect a theory—or grand conceptual scheme —to be fruitful in predictions and explanations, while keeping its assumptions as few and as general as possible.

Remember that a scientific "explanation" is neither an ultimate "reason why" from some inspired source nor a mere jumble of words describing observed behavior in technical terms. It is a linking of the observed behavior with some other well-known facts or with more general knowledge derived from observation. The greater the number and variety of the facts thus linked together the more satisfied we feel with our theory. As confidence in it increases we "explain" some facts by linking them to speculative guesses in our theory. Yet those guesses in turn are linked to experimental knowledge in our structure of theory; so it is much the same kind of "explaining," now vouched for by our belief in the validity of our theory.

As we build our theoretical treatment, we start with practical assumptions and simple concepts closely related to experiment; then we devise more general concepts to rule the simpler ones; and we aim finally at deducing our whole picture of nature from a few general concepts.

Above all, we value good theory for the sense of intellectual satisfaction it gives us—our feeling of confidence in our knowledge and pleasure in the compact way we can express it. There is an art in choosing theory so that it gives the strongest sense of consistent knowledge, and that is what we mean by understanding nature.

"If God were to hold out enclosed in His right hand all Truth, and in His left hand just the ever-active search for Truth, though with the condition that I should forever err therein, and should say to me: 'Choose!' I should humbly take His left hand and say: 'Father! Give me this one; absolute Truth is for Thee alone.'"

—G. E. Lessing, *Eine Duplik* (1778)